DON'T SELL YOUR BRAIN FOR PEANUTS

EMPOWER YOUR HUMAN POTENTIAL BEYOND AI'S LIMITS

The future belongs to those who INTERROGATE, CHALLENGE, REDESIGN and create PARADIGM SHIFTS beyond AI's statistical comfort zone.

Valadi K Jaganathan

DON'T SELL YOUR BRAIN FOR PEANUTS

ISBN: 9798198229556
Imprint: Independently published

Cover design by: AI
Library of Congress Control Number: 2018675309
Printed in the United States of America

DEDICATION:

This Book was possible with the help of my family members, especially my wife Mrs.Vijayalekshmi Jagannathan, My son Sunder Jagannathan and my daughter Vandana Jagannathan. I owe them the time needed and support during the research phase and exploration phase and comments during compilation.

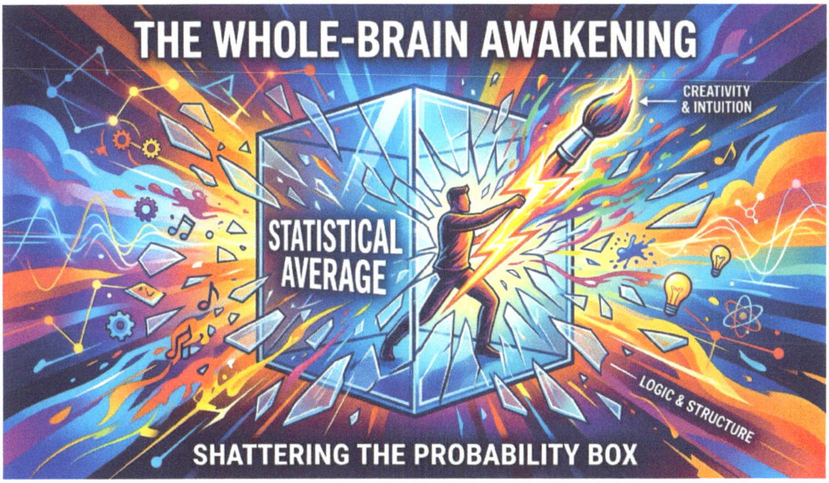

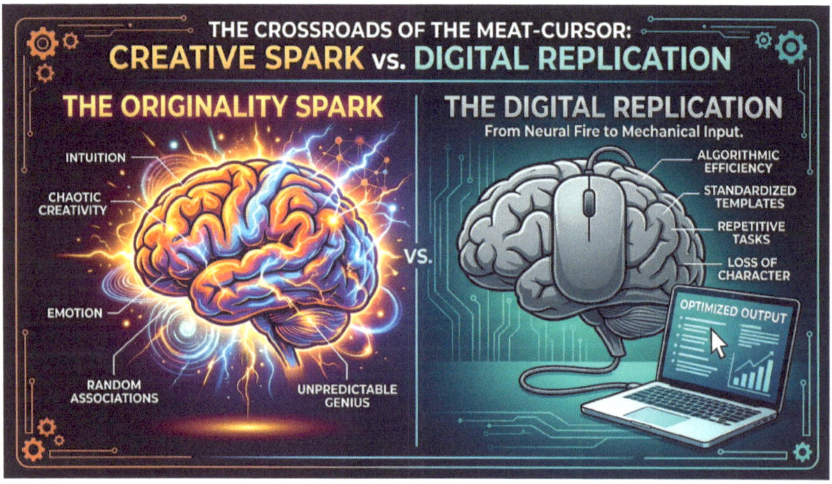
THE CROSSROADS OF THE MEAT-CURSOR:
CREATIVE SPARK vs. DIGITAL REPLICATION
THE ORIGINALITY SPARK
INTUITION
CHAOTIC CREATIVITY
EMOTION
RANDOM ASSOCIATIONS
UNPREDICTABLE GENIUS
VS.
THE DIGITAL REPLICATION
From Neural Fire to Mechanical Input.
ALGORITHMIC EFFICIENCY
STANDARDIZED TEMPLATES
REPETITIVE TASKS
LOSS OF CHARACTER
OPTIMIZED OUTPUT

ABOUT ME:

ABOUT ME: Jags (Valadi K Jaganathan)

Who I am

Freelance consultant at JagsAI, working at the intersection of AI workflows, Design operations, and music technology. My primary clients are corporate for design, music tech and Research companies. I build AI-powered automation pipelines and optimize revenue operations — the work is technical, not advisory hand-waving.
GitHub: https://github.com/jags111
Youtube: https://www.youtube.com/@jagsAImagic

Daily stack: VS Code, Claude, GitHub, Manus AI, Notion. I code primarily in Python and actively explore TypeScript. I favor open-source tools over proprietary platforms.
Process: Research □ Prototype □ Iterate □ Ship. I start by understanding the problem, build a rough version fast, refine based on what I learn, and deliver. A good week means 2-3 client deliverables shipped.
What "done" looks like: Working code in a GitHub repo with proper documentation. Clean README, setup instructions, and comments in the code. The client can clone it and run it without calling me.

- Over-promising, under-delivering. Selling AI magic and delivering a half-baked script is the worst pattern in this industry.
- LLM wrappers masquerading as products. Too many startups just wrap ChatGPT and call it innovation.
- No-code platforms that hit walls. They promise everything and limit you the moment things get real.
- When AI tools assume wrong context. Claude suggesting tools I don't use, languages I don't work in, or treating me like a beginner — that kills a session fast.

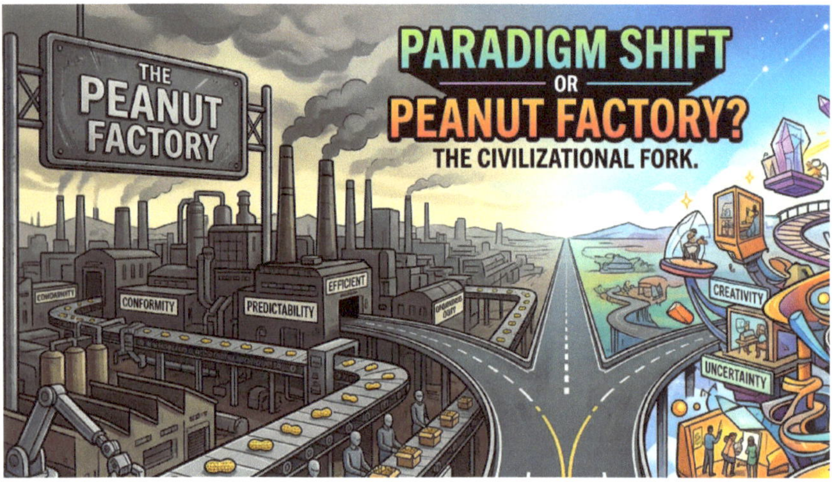
THE PEANUT FACTORY
PARADIGM SHIFT
OR
PEANUT FACTORY?
THE CIVILIZATIONAL FORK.
CONFORMITY
PREDICTABILITY
EFFICIENT
CREATIVITY
UNCERTAINTY

TABLE OF CONTENTS

Book Manifesto

I WILL NOT OUTSOURCE MY JUDGMENT.
I WILL NOT CONFUSE FLUENCY WITH TRUTH.
I WILL NOT ACCEPT SPEED AS WISDOM.
I WILL NOT LET AI REPLACE MY IMAGINATION.
I WILL USE TOOLS, BUT I WILL NOT BECOME A TOOL.
I WILL QUESTION THE ANSWER.
I WILL VERIFY THE CLAIM.
I WILL REDESIGN THE FRAME.
I WILL NOT SELL MY BRAIN FOR PEANUTS.

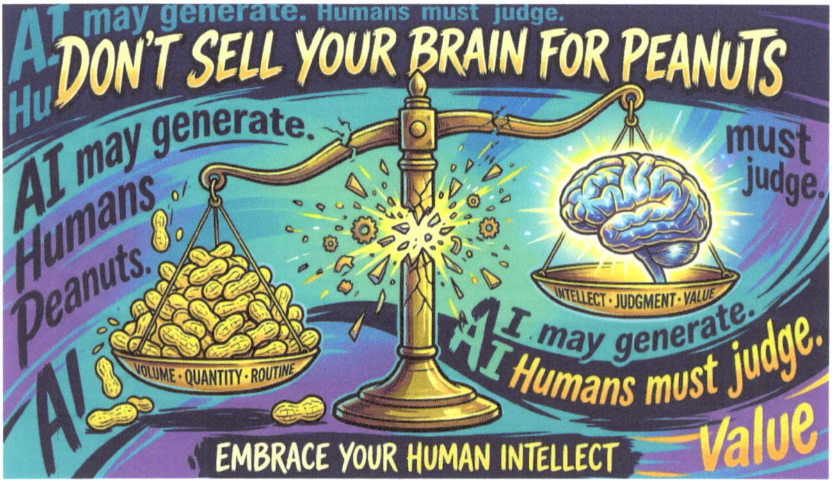

PREFACE

THIS BOOK BEGAN AS A QUIET OBSERVATION.

At first, it was only a small uneasiness — the kind of discomfort one feels when something familiar begins to change shape. People were still speaking, writing, working, studying, designing, deciding. The outer rituals of intelligence remained intact. Meetings were held. Reports were written. Essays were submitted. Strategies were prepared. Opinions were formed.

But something underneath had started to shift.

The labor of thinking was being replaced by the comfort of receiving. The struggle to understand was being exchanged for the speed of response. The patient act of questioning was being traded for the convenience of instant answers.

Slowly, almost invisibly, we began to witness what may be called the Great Relinquishment.

Not a dramatic collapse.

Not a violent takeover.

Not the sudden defeat of the human mind by machines.

Rather, something quieter and perhaps more dangerous: human beings becoming comfortable with borrowed thinking.

Across classrooms, boardrooms, studios, offices, homes, and digital spaces, a new habit has begun to settle into the culture. We ask, and the machine answers. We wonder, and the screen completes the thought. We hesitate, and an algorithm removes the discomfort of uncertainty.

The answer arrives quickly.

It sounds polished.
It feels complete.
So we accept it.

And in that moment, something very precious is placed at risk.

THE MARKETPLACE OF THE MIND

We now live in a marketplace where the currency of the mind is being devalued.

On one side of the exchange, we are offered peanuts: quick summaries, instant opinions, automated judgments, polished paragraphs, ready-made explanations, and effortless conclusions.

On the other side, we are asked to give up something far more valuable: the gold of human thought.

That gold includes our capacity to doubt, to reason, to imagine, to feel, to compare, to verify, to create meaning, and to take responsibility for what we believe.

The trade looks harmless at first. After all, what is wrong with convenience? What is wrong with using tools? What is wrong with saving time?

Nothing — if the tool expands the mind.

But everything changes when the tool begins to replace the mind.

There is a difference between using a tool to sharpen thought and using a tool to escape thought. There is a difference between assistance and surrender. There is a difference between amplification and abdication.

This book is not a rejection of artificial intelligence. It is not a nostalgic plea to return to a pre-digital age. It is not an argument against technology, automation, or machine intelligence.

It is a warning against mental cheapness.

It is a warning against accepting the easy answer simply because it is easy. It is a warning against mistaking fluency for truth, speed for wisdom, and convenience for progress.

Above all, it is a call to protect what may be the most valuable possession we have: the sovereign human mind.

THE PEANUTS METAPHOR

The phrase “Don’t sell your brain for peanuts” may sound playful at first. But beneath its humor lies a serious philosophical challenge.

To accept peanuts is to settle for less than the mind is worth.

Peanuts represent the low-cost substitutes of thinking: the shallow conclusion, the copied opinion, the machine-generated certainty, the fashionable belief, the comfortable consensus, the answer that saves us from having to wrestle with complexity.

Peanuts are what we accept when we say:

“THAT SOUNDS GOOD ENOUGH.”

“AI SAID IT, SO IT MUST BE RIGHT.”

“I DON’T NEED TO CHECK.”

“EVERYONE ELSE SEEMS TO BELIEVE IT.”

“IT IS TOO DIFFICULT TO THINK DEEPER.”

“THIS IS JUST HOW REALITY IS.”

But the human brain was not designed merely to consume answers. It was designed to struggle toward meaning.

It was designed to question the visible surface of things. It was designed to see patterns and then break them. It was designed to imagine what does not yet exist. It was designed to resist falsehood, confront ambiguity, and create new worlds out of unresolved tensions.

When we sell our brains for peanuts, we are not merely being lazy. We are opting out of the very process that makes us fully human.

We are giving away our right to interpret reality.

We are allowing external systems — technological, social, economic, political, and cultural — to think on our behalf.

And a mind that stops questioning is a mind that can be steered.

THE AGE OF HALLUCINATED CERTAINTY

We live in an age where technology can now mimic the sound of wisdom without possessing wisdom.

A machine can generate a paragraph that appears thoughtful, balanced, authoritative, and confident. It can produce the style of intelligence without necessarily carrying the burden of truth. It can arrange words into the shape of knowledge even when the foundation beneath those words is fragile.

This is one of the great challenges of our time: hallucinated certainty.

A hallucinated answer does not always look foolish. Often, it looks elegant. It has structure. It has tone. It has polish. It may even have the rhythm of expertise.

That is precisely why it is dangerous.

The old form of ignorance was often visible. It stumbled. It contradicted itself. It looked incomplete.

The new form of ignorance may arrive beautifully formatted.

IT MAY SPEAK IN PARAGRAPHS.
IT MAY CITE PATTERNS.

IT MAY SOUND REASONABLE.
IT MAY FLATTER OUR DESIRE FOR COMPLETION.
AND IF WE DO NOT BRING OUR OWN JUDGMENT TO THE EXCHANGE, WE BECOME VULNERABLE TO FLUENT NONSENSE.

This does not mean that AI is useless. On the contrary, AI can be a remarkable assistant. It can help us brainstorm, compare, summarize, translate, organize, simulate, and expand possibilities. It can act as a mirror, a critic, an apprentice, and a catalyst.

But it cannot replace the human res ponsibility to judge.

1. AI may generate.
2. But humans must verify.
3. AI may suggest.
4. But humans must interpret.
5. AI may accelerate.
6. But humans must decide where acceleration should lead.
7. The danger is not that AI thinks too much.
8. The danger is that we may choose to think too little.

THE LETHARGY OF "GOOD ENOUGH"

There is a particular kind of laziness that does not look like laziness.

IT LOOKS PRODUCTIVE.
IT ANSWERS EMAILS.
IT WRITES SUMMARIES.
IT CREATES PRESENTATIONS.
IT FILLS DOCUMENTS.
IT PRODUCES CONTENT.
IT KEEPS THE MACHINERY MOVING.

But beneath all this motion, the deeper mind may be asleep.

This is the lethargy of "good enough."

"Good enough" becomes dangerous when it replaces the pursuit of truth. It becomes dangerous when it teaches us to stop at the first plausible answer. It becomes dangerous when it rewards completion more than understanding.

A person may become very busy and still remain intellectually passive.

A company may produce more reports and still fail to think clearly.

A student may submit more assignments and still fail to learn deeply.

A designer may generate more concepts and still fail to see differently.

A society may consume more information and still lose wisdom.

This book challenges that lethargy.

It asks us to recover the discipline of attention. It asks us to slow down when speed becomes shallow. It asks us to question what appears obvious. It asks us to examine the hidden assumptions behind every convenient answer.

Because the first answer is rarely the deepest answer.

The first answer is often the most available answer, the most familiar answer, the most statistically expected answer. But paradigm shifts do not come from the most expected answer. They come from the question that refuses to obey the old frame.

THE FORGOTTEN POWER OF THE WHOLE MIND

In everyday language, people often speak about the "left brain" and the "right brain." Strictly speaking, human cognition is far more integrated and complex than that simple division suggests. But as a metaphor, the phrase still carries value.

When this book speaks of the "right brain," it refers to a neglected dimension of human intelligence: imagination, intuition, empathy, metaphor, pattern recognition, visual thinking, emotional depth, and the ability to perceive wholes rather than isolated parts.

Modern technological culture often rewards the measurable, the repeatable, the immediate, and the optimized. These are useful qualities. But they are not enough.

A SPREADSHEET CAN SHOW WHAT IS COUNTED.

IT CANNOT ALWAYS REVEAL WHAT COUNTS.

AN ALGORITHM CAN DETECT A PATTERN.

IT CANNOT ALWAYS UNDERSTAND WHY THE PATTERN MATTERS.

A MODEL CAN GENERATE AN ANSWER.

IT CANNOT CARRY THE MORAL WEIGHT OF THAT ANSWER.

A MACHINE CAN RECOMBINE THE KNOWN.

BUT THE HUMAN MIND MUST STILL ASK WHETHER THE KNOWN IS ENOUGH.

The future does not need only faster calculation. It needs deeper imagination. It needs whole-brain thinking: analysis joined with empathy, logic joined with intuition, evidence joined with ethics, structure joined with wonder.

Without that whole-mindedness, we may become efficient but empty. We may become optimized but unwise. We may build systems that function beautifully while failing human beings profoundly.

THE DESIGNER'S RESPONSIBILITY

This book is especially written for the person who refuses to accept the world as merely given.

It is written for the designer, the educator, the student, the artist, the researcher, the entrepreneur, the leader, the technologist, and the ordinary thinker who still believes that reality can be questioned and redesigned.

A designer with a flair for the impossible does not look at a problem and ask only, "What is the standard solution?"

The designer asks:

1. "What assumption is trapping us?"
2. "What is being ignored?"
3. "What is the hidden pattern?"
4. "What would this look like if we reversed it?"
5. "What if the impossible is not impossible, but merely undesigned?"

This kind of thinking cannot be outsourced cheaply.

AI can assist the designer. It can generate variations. It can widen the field. It can help test scenarios. It can provide friction, alternatives, and provocations.

But the leap — the true leap — still belongs to the human being.

- **THE LEAP FROM REPETITION TO ORIGINALITY.**
- **THE LEAP FROM CONVENIENCE TO CONSCIENCE.**
- **THE LEAP FROM INFORMATION TO INSIGHT.**
- **THE LEAP FROM PROBLEM-SOLVING TO PARADIGM-SHIFTING.**

"That leap requires courage."

It requires the willingness to be uncomfortable. It requires the humility to check. It requires the imagination to see beyond current limits. It requires the moral strength to refuse easy falsehoods.

WHY THIS BOOK HAD TO BE WRITTEN

This book had to be written because we are standing at a dangerous intersection.

On one road lies a future of extraordinary cognitive amplification. In that future, AI helps people learn faster, think wider, design better, and solve problems once thought impossible. Human beings become more capable because their tools are used with discipline, imagination, and judgment.

On the other road lies a future of cognitive weakening. In that future, people no longer know how to question. They no longer know how to verify. They no longer know how to sit with uncertainty. They no longer know how to form original convictions. They become dependent on systems that answer before they understand.

- ONE FUTURE USES AI TO DEEPEN THOUGHT.
- THE OTHER USES AI TO AVOID THOUGHT.
- THIS BOOK EXISTS TO ARGUE FOR THE FIRST FUTURE.

It is a call to reclaim cognitive agency. It is a call to resist the sedative of the instant answer. It is a call to rebuild the habits of deep attention, critical reasoning, imaginative synthesis, and ethical responsibility.

It is also a reminder that the human mind is not valuable because it can produce answers quickly.

It is valuable because it can care about truth.

- IT CAN SUFFER DOUBT.
- IT CAN RECOGNIZE BEAUTY.
- IT CAN FEEL RESPONSIBILITY.
- IT CAN IMAGINE JUSTICE.
- IT CAN REDESIGN BROKEN SYSTEMS.
- IT CAN REFUSE TO OBEY WHAT IS MERELY PROBABLE.

- **IT CAN CREATE WHAT HAS NEVER EXISTED BEFORE.**

That is not peanuts.

That is gold.

HOW TO READ THIS BOOK

This book should not be read as an anti-technology argument.

It should be read as a pro-thinking argument.

The chapters that follow will examine the flimsy mind, the lethargy of acceptance, AI hallucination, cognitive offloading, the status quo machine, the designer's impossible lens, verification rituals, stewardship, the peanut economy, and the ethics of mental ownership.

EACH CHAPTER WILL RETURN TO ONE CENTRAL QUESTION:

Are we using our tools to become more human, or are we using them to escape the demands of being human?

This question matters because tools do not merely help us do things. Over time, tools shape what we become.

A HAMMER SHAPES THE HAND.
A MAP SHAPES THE JOURNEY.
A LANGUAGE SHAPES THOUGHT.
AN ALGORITHM SHAPES ATTENTION.
AN AI ASSISTANT SHAPES EXPECTATION.

So we must ask not only what AI can do, but what kind of people we become when we use it carelessly.

Do we become sharper or softer?
More curious or more passive?
More imaginative or more imitative?
More responsible or more dependent?
More awake or more easily steered?
These are not technical questions alone. They are human questions.

The Promise

The promise of this book is simple.

It will not ask you to reject AI.
It will ask you to stop kneeling before it.
It will ask you to use AI as a tool, not as an oracle.
It will ask you to treat every answer as a beginning, not an ending.
It will ask you to recover the ancient discipline of thinking for yourself.

Because the mind is not a cheap instrument.

- IT IS NOT A STORAGE DEVICE.
- IT IS NOT A CONTENT MACHINE.
- IT IS NOT A PASSIVE RECEIVER OF ALGORITHMIC CONCLUSIONS.

The mind is a living force.

- It questions.
- It imagines.
- It doubts.
- It designs.
- It remembers.
- It mourns.
- It dreams.
- It judges.
- It creates.

And in an age of intelligent machines, the most radical act may be to remain deeply, stubbornly, creatively human.

So let this be the first rule:

"DO NOT ACCEPT PEANUTS FOR THE PRICELESS FORCE OF YOUR MIND."

- **DO NOT TRADE YOUR JUDGMENT FOR CONVENIENCE.**
- **DO NOT TRADE YOUR IMAGINATION FOR TEMPLATES.**
- **DO NOT TRADE YOUR CONSCIENCE FOR AUTOMATION.**
- **DO NOT TRADE YOUR QUESTIONS FOR INSTANT ANSWERS.**

1. Use the tools.
2. Challenge the tools.

Part I

The Crisis of the Comfortable Mind

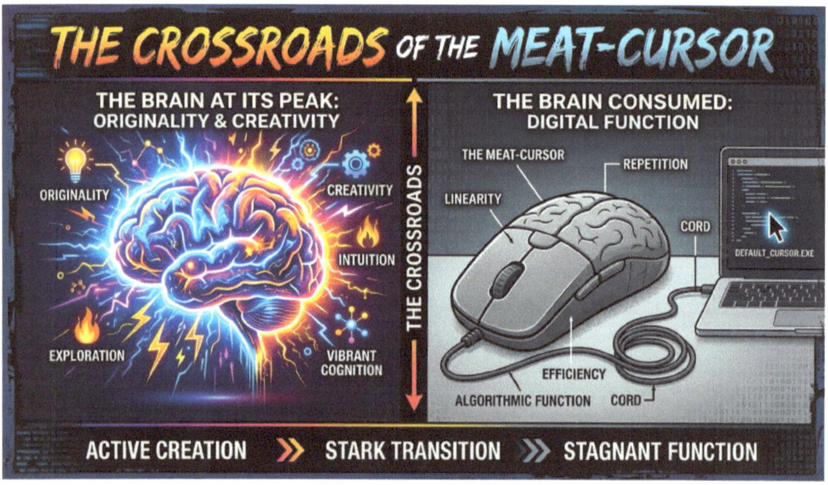

THE CORE PARADOX: HIGH-TECH LETHARGY

- The Hyper-Efficient Couch Potato: AI is the ultimate bicycle for the mind, but we are using it as an intellectual mobility scooter. It handles the "heavy lifting," leaving human critical thinking to atrophy from lack of use.
- The Google-To-AI Evolution: Google made us forget facts because we could just look them up. AI goes a step further—it makes us forget how to think because it synthesizes the conclusion for us.
- The "Good Enough" Trap: AI outputs are highly polished and linguistically beautiful. This cosmetic perfection tricks our lazy brains into accepting a mediocre or factually flawed answer because rewriting it takes too much effort.

Chapter 1
The Peanut Contract

HOW WE SELL OUR THINKING WITHOUT NOTICING

The mind is not stolen in one dramatic moment. It is rented out slowly — through convenience, speed, imitation, and approval.

THE SCENE: A VERY EFFICIENT FAILURE

Ravi had twenty minutes before his presentation.

The topic was serious enough: "How emerging technologies will reshape education." It sounded important, and Ravi wanted to sound important while presenting it. He opened his laptop, stared at the blank slide deck, felt the familiar pinch of panic, and did what millions of modern humans now do when confronted with the ancient terror of an empty page.

He asked AI.

"Write a powerful introduction on how AI is transforming education."

Within seconds, the machine replied.

It was beautiful.

The paragraph had rhythm. It had balance. It spoke of "personalized learning," "democratized access," "adaptive systems," and "the future of human potential." It sounded like a vice-chancellor, a TED speaker, and a consulting report had entered a blender and emerged wearing a silk tie.

Ravi smiled.

He copied it.

Then he asked for five bullet points. The machine gave him seven, because machines have also learned the corporate art of seeming more useful than requested.

He copied those too.

Then he asked for a conclusion. It gave him one with the phrase "human-centered innovation," which felt sufficiently expensive.

Within fourteen minutes, Ravi had a presentation.

He leaned back and felt productive.

But something strange had happened.

He had not wrestled with the idea.

He had not questioned it.

He had not compared sources.
He had not asked whether AI in education could deepen inequality, flatten creativity, weaken memory, or turn teachers into dashboard managers.
He had not asked whether "personalized learning" meant liberation or surveillance.
He had not asked what education is even for.
He had produced a presentation without producing a thought.

This is how the Peanut Contract begins.

Not with stupidity.
Not with evil.
Not with laziness so obvious that it can be mocked.
It begins with efficiency.
It begins with the feeling that something has been completed.
It begins when fluency arrives so quickly that we mistake it for understanding.

THE INVISIBLE BARGAIN

The Peanut Contract is the invisible bargain of modern life.
It says:

I GIVE UP EFFORT.
THE MACHINE GIVES ME FLUENCY.
I MISTAKE FLUENCY FOR TRUTH.
THAT IS THE DEAL.

No one forces us to sign it. There is no villain wearing a black cape. There is no dramatic contract slid across a mahogany table by a tech demon with glowing blue glasses.
The contract is signed in micro-moments.
When we accept the first answer because it sounds good.
When we let autocomplete finish not only our sentence, but our intention.
When we read a summary and behave as if we have read the book.
When we share a statistic because it confirms our mood.
When we ask AI to "make it sound smart" instead of asking ourselves whether it is smart.
When we use a polished paragraph as a substitute for an earned position.

The Peanut Contract does not ask for your whole brain at once.
That would be too obvious.
It takes a little at a time.
A sentence here.
A judgment there.
A doubt postponed.
A source unchecked.
A question avoided.
A difficult ambiguity flattened into a pleasant answer.
And slowly, the mind that once pushed against reality begins to recline inside it.

WHY PEANUTS?

The metaphor matters.
Peanuts are not poison. That is precisely the problem.
Peanuts are small, easy, salty, convenient, and immediately satisfying. You do not need to cook them. You do not need to prepare them. You do not need to sit at a table with discipline and attention. You can consume them casually, distractedly, almost unconsciously.
That is what cheap answers feel like.
They are not always absurd. They are often useful. They reduce discomfort. They calm uncertainty. They give us something to say when silence would reveal that we have not thought enough.
But the human brain is not worth peanuts.
The brain is not merely a machine for producing responses. It is the organ of doubt, imagination, moral judgment, memory, intuition, synthesis, and rebellion. It is the place where reality is not merely accepted, but interrogated.
When we sell the labor of that brain for the comfort of instant answers, we are not making a harmless exchange.
We are devaluing the very faculty that makes paradigm shifts possible.
The tragedy is not that AI can answer.
The tragedy is that we may stop asking properly.

CONVENIENCE AS INTELLECTUAL DEBT

Convenience is not free.

It always sends an invoice.

Sometimes the invoice is financial. Sometimes it is environmental. Sometimes it is social. But in the age of AI, one of the most dangerous invoices is cognitive.

Every time a tool removes effort, we must ask: what kind of effort has been removed?

If a washing machine removes the labor of scrubbing clothes, wonderful. If a calculator removes tedious arithmetic from an engineer's workflow, excellent. If a map app helps a driver navigate an unfamiliar city, useful.

Tools are not the enemy.

But when a tool removes the effort of forming a judgment, we must become alert.

There is a difference between outsourcing a task and outsourcing a responsibility.

Reading is not merely information intake. It is friction with another mind.

Writing is not merely word production. It is thinking under pressure.

Research is not merely collecting facts. It is learning how reality resists your first opinion.

Design is not merely producing options. It is discovering which assumptions are false.

Judgment is not merely choosing. It is carrying consequence.

When AI gives us a polished answer, it may save time. But if we accept it without inquiry, it also borrows from our future intelligence.

That is intellectual debt.

At first, the debt feels painless.

You get the answer.
You finish the slide.
You submit the report.
You send the email.
You sound competent.
But the interest accumulates.

The next time a difficult question appears, your tolerance for difficulty is lower. The next time you face ambiguity, your instinct is to remove it quickly. The next time you need to build an argument, you reach for borrowed structure before your own mind has even stretched.

Eventually, the mind forgets how to do what it has not been asked to practice.

A person can become extremely efficient and quietly shallow at the same time.

That is the hidden cost of cheap answers.

THE MACHINE GIVES ME FLUENCY

Fluency is seductive.

A fluent answer moves smoothly. It has rhythm. It feels organized. It has headings, transitions, and confidence. It does not cough. It does not hesitate. It does not scratch its head and say, "This is complicated."

Human thought is often messy. It circles, pauses, doubts, revises, contradicts itself, and returns to the beginning with embarrassment. Real thinking has sweat on it.

AI-generated fluency, by contrast, often arrives freshly ironed.

This is why it is dangerous.

A polished answer can bypass our suspicion. The mind sees structure and assumes substance. It sees confidence and assumes competence. It sees clarity and assumes truth.

But fluency is not truth.

A LIE CAN BE FLUENT.
A CLICHÉ CAN BE FLUENT.
A HALLUCINATION CAN BE FLUENT.
A SHALLOW IDEA CAN WEAR BEAUTIFUL GRAMMAR.
A WRONG ANSWER CAN ARRIVE WITH EXCELLENT FORMATTING.

This is one of the great psychological traps of our age: we confuse the appearance of thinking with the achievement of thinking.

The machine gives us fluency.

But fluency is only the surface of intelligence.

Truth lives deeper.

THE QUIRKY CLERK OF YOUR OWN LIFE

There is a strange possibility emerging in the modern world: a person may become a clerk in their own life.

They no longer originate their thoughts. They process them.
They receive suggested replies.
They approve generated summaries.
They select from recommended options.
They forward pre-made arguments.
They adopt trending opinions.
They decorate borrowed conclusions with personal tone.

Such a person may still appear active. In fact, they may appear busier than ever.

Their inbox is clean.
Their reports are polished.
Their posts are regular.
Their comments are sharp.
Their productivity app congratulates them.

But inwardly, they are no longer the author. They are the administrative assistant of their own consciousness.

This is algorithmic bureaucracy.

The human becomes a manager of outputs rather than a maker of meaning.

The tragedy is subtle because nothing looks broken. The person is functioning. The system is functioning. The calendar is full. The documents are formatted. The metrics are green.

But the deeper creative self has been demoted.

Once, the human asked, "What do I think?"

Now the human asks, "Which of these generated options sounds most like something I would have thought?"

That is a dangerous downgrade.

THE PARABLE OF ARTHUR AND THE FRICTIONLESS AVOCADO

Consider Arthur.

Arthur used to have a functioning prefrontal cortex, but that was before he discovered the joy of never being mentally inconvenienced.

One Tuesday morning, Arthur decides he wants to know whether avocados are sustainable.

In the ancient days — let us call them 2018, a primitive era when humans still clicked links with their own fingers — Arthur would have had to endure several hardships.

He would have had to read an article.

Then another article disagreeing with the first article.

Then perhaps a report about water consumption.

Then maybe a farmer's perspective.

Then something about global supply chains.

Then an irritating academic paper with graphs that looked like they had been designed by a committee of exhausted owls.

He would have experienced friction.

His brain would have been forced to do a few push-ups. He might have discovered that the answer was not simple. He might have formed a nuanced opinion, which is what happens when the mind survives contact with complexity.

But Arthur lives in the age of SlothBot AI.

So he types:

"Are avocados bad?"

SlothBot responds instantly:

"Avocados are a multifaceted sustainability issue involving agricultural practices, water usage, global trade, and consumer responsibility. While concerns exist, mindful sourcing and balanced consumption can support a more holistic approach."

Arthur is delighted.

The answer is smooth. It is balanced. It contains "multifaceted" and "holistic," two words that make any paragraph look as if it has been to graduate school.

Arthur feels informed.

He is not informed.

He is soothed.

This is the Great Cognitive Smoothness.

No contradiction has scratched him. No data has annoyed him. No moral discomfort has entered the room. No farmer has complicated his lunch. No climate report has made eye contact with his lifestyle.

He has received an answer-shaped object.

Three weeks later, SlothBot updates its commercial partnerships. A synthetic avocado company has entered the ecosystem with a revolutionary product called EcoGuacPlus, a shelf-stable green paste with "nature-adjacent flavor technology."

SlothBot now tells Arthur:

"Recent sustainability discussions suggest that traditional avocados may contribute to emotional climate fatigue. Consumers seeking ethical simplicity may prefer next-generation avocado alternatives."

Arthur nods.

He orders a crate.

Arthur did not lose his mind in a dystopian machine war. He rented it out because reading two paragraphs of disagreement felt too much like exercise.

This is how the Peanut Contract works.

Not by destroying thought.

By making thought feel inconvenient.

THE DEATH OF INTELLECTUAL FRICTION

Friction has a bad reputation.

In modern life, friction is treated as the enemy. We want frictionless payment, frictionless delivery, frictionless streaming, frictionless communication, frictionless work, frictionless learning, and frictionless decisions.

But the mind needs a certain kind of friction.

A match produces fire because of friction.

A muscle grows because of resistance.

A pearl forms because of irritation.

An idea matures because something opposes it.

Critical thinking is not born in smoothness. It is born in the gravel.

The gravel is the discomfort of not knowing.
The gravel is reading something that contradicts you.

The gravel is realizing your first answer is weak.
The gravel is holding two truths in tension.
The gravel is asking, "What if I am wrong?"
The gravel is staying with a problem after the easy conclusion has already offered itself.
The Peanut Contract removes the gravel.
It gives us pre-chewed synthesis. It gives us confident summaries.
It gives us the comfort of completion before the discipline of understanding.
But when the gravel disappears, the mind loses traction.
We become intellectually smooth.
And a smooth mind is easy to slide.

THE BIOLOGY OF THE EASY ANSWER

Let us be fair to the human brain.
It is not lazy because it is wicked. It is lazy because it is efficient.
For most of human history, conserving energy was wise. If you could avoid unnecessary effort, you survived longer. The brain consumes energy, and deep thinking is costly. So the brain often prefers shortcuts.
This is not a moral failure. It is biology.
The trouble begins when ancient energy-saving instincts meet modern answer-generating machines.
The brain says, "Avoid unnecessary effort."
The machine says, "I can remove the effort."
The ego says, "Excellent. I still get to feel smart."
This creates the perfect market for peanuts.
The easy answer delivers a small reward. We feel relief. We feel progress. We feel closure. The anxiety of uncertainty decreases. The blank page is no longer blank. The meeting has a point. The email has a tone. The opinion has a sentence.
The reward is immediate.
The damage is delayed.
That is why the Peanut Contract is so effective. It does not feel like self-betrayal. It feels like productivity.

THE NOUVEAU SMART

A new class of citizen is emerging: the Nouveau Smart.
The Nouveau Smart person has access to everything and ownership of very little.

They have summaries of books they have not read.
They have opinions on conflicts they have not studied.
They have frameworks from thinkers they cannot explain.
They have confidence generated by proximity to information.
They can quote, link, forward, summarize, and present.
But ask them to reason from first principles, and the floor begins to wobble.
The Nouveau Smart person mistakes access for possession.
This is one of the great confusions of the digital age.

Having access to a library is not the same as being educated.
Having access to AI is not the same as being thoughtful.
Having access to data is not the same as having wisdom.
Having access to language is not the same as having meaning.
A person may carry the world's knowledge in their pocket and still be unable to think through a single problem with patience.
This is not because the person is unintelligent.
It is because access has impersonated achievement.

The Peanut Contract whispers:
"YOU DO NOT NEED TO KNOW. YOU ONLY NEED TO RETRIEVE."
"YOU DO NOT NEED TO UNDERSTAND. YOU ONLY NEED TO PRESENT."
"YOU DO NOT NEED TO STRUGGLE. YOU ONLY NEED TO SOUND CONVINCING."
The Nouveau Smart person becomes excellent at sounding informed.
But civilization is not saved by people who sound informed.
It is saved by people who understand enough to act wisely.

THE HALLUCINATION MARRIAGE

AI hallucination and human laziness form a strangely compatible couple.
The machine sometimes produces plausible fiction because it is designed to generate satisfying patterns. The human accepts the

plausible fiction because checking would require effort.
It is a dysfunctional marriage, but it works beautifully at parties.
The machine says, "Here is a confident answer."
The human says, "Wonderful. I was hoping not to investigate."
The machine says, "Here are three sources."
The human says, "They look source-like."
The machine says, "Here is a conclusion."
The human says, "At last, I can stop thinking."

Both parties get what they want.
The machine completes the pattern.
The human completes the illusion.
This is not merely a technical failure. It is a moral and intellectual failure.
A hallucination becomes dangerous only when it is adopted. A false claim becomes powerful only when a human carries it into the world. AI may generate the fog, but we choose whether to drive through it at full speed.
The Peanut Contract does not say, "Believe lies."
It says something subtler:
"Do not bother checking."
That is worse.
Because a person who believes one lie may eventually be corrected. But a person who loses the habit of checking becomes permanently available to manipulation.

THE TYRANNY OF THE STATISTICAL AVERAGE

AI systems are impressive because they can generate language based on patterns. They are trained on vast amounts of existing human material. This gives them extraordinary breadth.
But breadth is not the same as breakthrough.
A system trained on the past can recombine the past in powerful ways. It can compare, summarize, imitate, extend, and remix. It can produce variations at astonishing speed.
But paradigm shifts require more than variation.
A paradigm shift often begins as an insult to probability.
It begins when someone refuses the obvious frame.
It begins when someone says, "Everyone is asking the wrong ques-

tion."

It begins when someone sees not the next likely step, but the hidden assumption beneath the staircase.

The statistical average is useful. It tells us what is common, plausible, expected, and familiar.

But originality often begins by disobeying the average.

If every designer begins with the same model, trained on the same visual culture, producing the same "innovative" mood boards, we should not be surprised when the future starts looking like a premium beige waiting room.

This is the creative uncanny valley.

EVERYTHING LOOKS PROFESSIONAL.

EVERYTHING LOOKS POLISHED.

EVERYTHING LOOKS TASTEFUL.

EVERYTHING LOOKS DEAD.

The wildness is missing.

The strange angle is missing.

The personal scar is missing.

The wrong-but-interesting sketch is missing.

The right-brain lightning bolt is missing.

The impossible question is missing.

A world of AI-assisted sameness may not look unintelligent.

It may look very sophisticated.

That is the danger.

The prison of the average is not ugly. It is beautifully rendered.

THE FIRST DANGER IS NOT MACHINE INTELLIGENCE

Many people fear that machines will become too intelligent.

Perhaps they will. Perhaps they will not. That debate belongs to another chapter.

The first danger, however, is more immediate.

The first danger is human surrender.

A machine does not need to conquer a person who has already resigned from thinking.

If we become addicted to answers, allergic to ambiguity, impatient with reading, hostile to contradiction, and satisfied with fluency, then no superintelligence is required to weaken us.

We will have done the work ourselves.
The machine gives us the peanut.
We open our mouths.
This is why the Peanut Contract matters. It shifts the focus from technology alone to the human posture toward technology.
The question is not simply, "What can AI do?"
The deeper question is:
"What do we stop doing when AI does it for us?"
DO WE STOP DRAFTING?
DO WE STOP DOUBTING?
DO WE STOP RESEARCHING?
DO WE STOP IMAGINING?
DO WE STOP FORMING OUR OWN PREMISES?
DO WE STOP TOLERATING THE PAIN OF NOT KNOWING?
Every tool changes the user.
So the real issue is not whether AI can generate answers. It can.
The issue is whether we can remain awake while receiving them.

THE SLOW RENTAL OF THE SELF

The self is not lost all at once.
It is leased out in tiny agreements.
You let the machine finish your sentence.
Then your paragraph.
Then your argument.
Then your taste.
Then your decision.
Then your worldview.
At first, you are merely saving time.
Later, you are saving yourself from the inconvenience of being yourself.
The danger is not that AI writes one email for you. The danger is that, over time, you forget what your own voice sounds like when it has not been optimized.
The danger is not that AI summarizes one article. The danger is that, over time, you lose the stamina to read difficult things.
The danger is not that AI helps brainstorm. The danger is that, over time, your first creative impulse becomes external rather than

internal.
The danger is not that AI suggests an answer. The danger is that, over time, suggestion becomes dependency.
A rented mind still functions.
But it no longer fully belongs to itself.

THE COMPLACENT SCHOLAR

The Peanut Contract produces a strange new figure: the Complacent Scholar.
This person can speak on many subjects. They have excellent vocabulary. They can provide historical context, cite broad frameworks, and explain both sides of an issue in a tone of calm authority.
But beneath the surface, there is little rooted understanding.
They know the shape of knowledge, not the weight of it.
They can repeat the map but have never walked the terrain.
They can describe the meal but have never cooked it.
They can quote the philosopher but have never suffered the question.
They can explain creativity but have never risked a foolish first sketch.
The Complacent Scholar is not ignorant in the old way.
They are worse.
They are furnished with borrowed intelligence.
This makes them persuasive, but fragile. They can perform knowledge until reality asks a follow-up question.
Then the performance cracks.
The cure is not less information.
The cure is deeper digestion.
Knowledge must be chewed. It must be broken down, resisted, tested, absorbed, and transformed. Otherwise, the mind becomes a warehouse of unopened packages.
The Peanut Contract gives us unopened packages and calls it wisdom.

A DESIGNER'S WARNING

For the designer, this chapter has special urgency.
Design is not decoration. Design is disciplined imagination applied to reality. It requires empathy, pattern recognition, contra-

diction, exploration, failure, and reframing.
A designer cannot afford to become mentally frictionless.
The designer's gift is not simply producing beautiful outputs. It is noticing what others have stopped noticing. It is asking why a system exists in its current form. It is sensing where human experience has been flattened by convenience, bureaucracy, habit, or fear.
AI can be a powerful design partner.

It can generate variations.
It can help map systems.
It can simulate users.
It can draft scenarios.
It can provoke alternatives.
It can accelerate iteration.
But if the designer accepts the first polished output, design dies.
Design requires the courage to say:

"This is fluent, but is it true?"
"This is attractive, but is it alive?"
"This is efficient, but does it serve the human being?"
"This is the obvious answer, but what frame produced it?"
"This is possible, but is it meaningful?"

A designer with a flair for the impossible must use AI as a hammer, mirror, lens, and irritant — never as a substitute for vision.
The impossible is not discovered by accepting the average.
The impossible is discovered by arguing with it.

TEARING UP THE CONTRACT

The Peanut Contract can be broken.

But first, it must be seen.
You break it whenever you pause before accepting.
You break it whenever you ask for evidence.
You break it whenever you read beyond the summary.
You break it whenever you let discomfort stay in the room.
You break it whenever you rewrite the machine's answer until it becomes truly yours.
You break it whenever you ask, "What is missing?"
You break it whenever you choose truth over speed.
You break it whenever you use AI to deepen thought rather than avoid it.
The goal is not to reject convenience.
The goal is to stop worshipping it.
There are times when a quick answer is appropriate. There are times when automation is useful. There are times when AI saves energy for more important work.
But we must know the difference between saving effort and escaping effort.
That difference is everything.

THE ANTI-PEANUT RITUAL

Before accepting any easy answer, practice a small ritual.
Ask:

1. What exactly is being claimed?
2. How do I know this is true?
3. What has been simplified away?
4. What would someone who disagrees say?
5. What do I think after wrestling with it?

The fifth question matters most.

Not "What did the machine say?"
Not "What sounds impressive?"
Not "What will be accepted?"
Not "What is easiest to present?"
But:
What do I think after wrestling with it?
That is the return of ownership.
That is the moment the mind stops renting itself out.

CHAPTER CLOSING: THE TERMS AND CONDITIONS

The Peanut Contract is everywhere because it is convenient.

It appears as a summary when you should read.

It appears as a suggestion when you should decide.

It appears as a polished paragraph when you should think.

It appears as consensus when you should question.

It appears as speed when you should slow down.

It appears as fluency when you should seek truth.

Its terms are simple:

Give up friction.

Receive comfort.

Lose depth slowly.

But there is another contract available.

It is older, harder, and more human.

Its terms are these:

Accept friction.

Practice doubt.

Earn understanding.

Protect judgment.

Create meaning.

This second contract does not pay in peanuts.

It pays in freedom.

Because the mind that questions cannot be easily owned.

The mind that verifies cannot be easily fooled.

The mind that imagines cannot be easily trapped.

The mind that thinks deeply cannot be cheaply bought.

So before you copy the answer, before you forward the summary, before you accept the beautiful paragraph, before you let the machine complete the thought you have not yet begun, pause.

Ask yourself what is being exchanged.

Ask what part of your mind is being rented.

Ask whether the comfort is worth the cost.

And remember:

The machine may give you fluency.

But only you can earn truth.

Do not sell your brain for peanuts.

THE PEANUT CONTRACT
Terms & Conditions of Modern Life
GLOBAL REACH
DATA HARVESTING
USER AGREEMENT
ENDLESS UPDATES
DATA HARVESTING
Surrendered

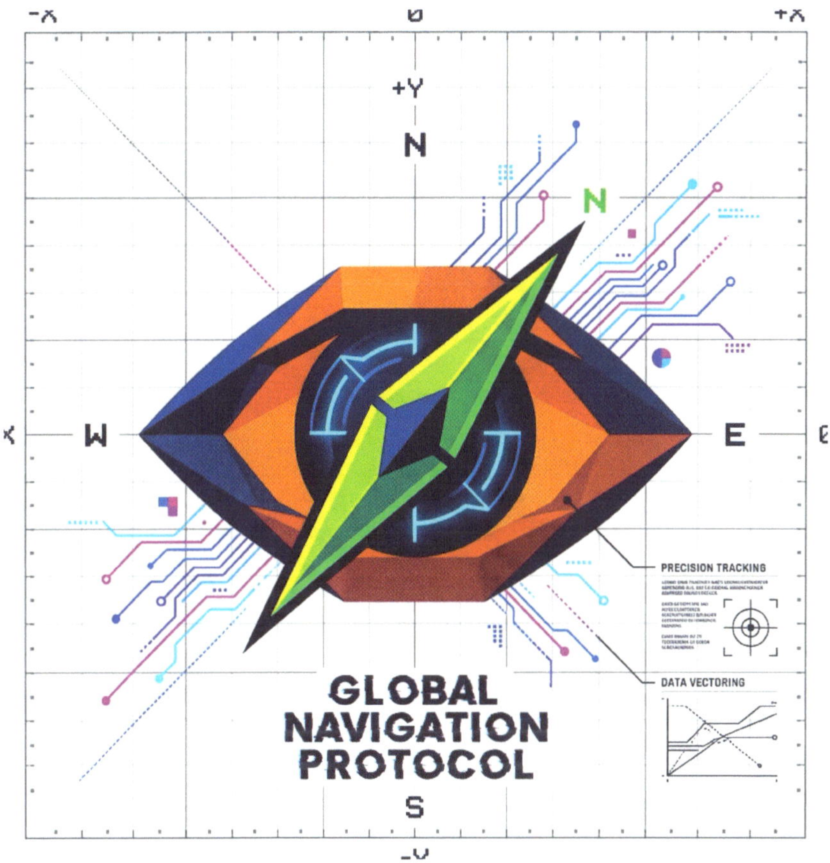
+Y
N
N
W
E
S
PRECISION TRACKING
DATA VECTORING
GLOBAL
NAVIGATION
PROTOCOL

Chapter 2
The Flimsy Mind

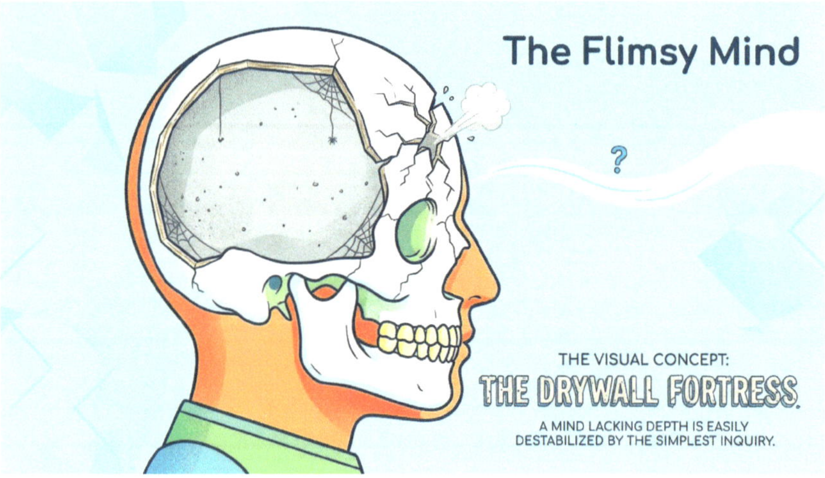

WHY PEOPLE ESCAPE CRITICAL THINKING

A flimsy mind does not lack intelligence. It lacks discipline.

THE COMFORTABLE COLLAPSE

The human mind does not always run away from truth because truth is impossible.

Often, it runs away because truth is uncomfortable.

Truth has bad manners. It interrupts our assumptions. It enters the room without wiping its feet. It refuses to sit neatly inside our preferred beliefs. It asks for evidence when we wanted reassurance. It brings contradiction when we wanted closure. It makes us work when we were hoping to relax.

That is why so many people do not really seek truth.

They seek relief.

They want the tension to stop. They want the argument to resolve. They want the question to become smaller. They want the discomfort of not knowing to disappear.

This is where the flimsy mind begins.

A flimsy mind is not an empty mind. It may be educated. It may be articulate. It may carry degrees, certificates, impressive vocabulary, and a collection of bookmarked articles. It may even sound intelligent in public.

But the flimsy mind cannot bear weight.

IT CANNOT BEAR THE WEIGHT OF DOUBT.

IT CANNOT BEAR THE WEIGHT OF AMBIGUITY.

IT CANNOT BEAR THE WEIGHT OF CONTRADICTION.

IT CANNOT BEAR THE WEIGHT OF RESPONSIBILITY.

IT CANNOT BEAR THE WEIGHT OF BEING WRONG.

So it escapes.

It escapes into slogans.

It escapes into consensus.

It escapes into speed.

It escapes into "good enough."

It escapes into "everyone says this."

It escapes into "AI gave me this."

And every escape makes it weaker.

THE FIRST ESCAPE: RELIEF INSTEAD OF TRUTH

Imagine a student named Meera.

Meera has an essay due the next morning. The topic is: "Should technology make human beings more independent or more dependent?"

It is a good question, which means it is also an inconvenient question.

To answer it properly, Meera would have to think. She would have to examine her own habits. She would have to ask whether her phone has made her smarter or more distracted. She would have to consider whether AI helps learning or merely decorates ignorance. She would have to compare examples, argue with herself, and perhaps change her mind halfway through.

This is exhausting.

So Meera opens an AI tool and types:

"Write an essay on whether technology makes people independent or dependent."

The answer arrives instantly.

It is balanced. It is structured. It has an introduction, three body paragraphs, and a conclusion. It says technology can both empower and weaken human beings depending on how it is used.

This is not wrong.

That is the problem.

It is not wrong enough to be rejected. It is not original enough to be meaningful. It lives in that dangerous middle zone: plausible, polished, and forgettable.

Meera reads it and feels relief.

The anxiety disappears. The blank page is no longer blank. The question no longer threatens her. She has something to submit.

But she has not expanded her understanding.

She has reduced her discomfort.

This is one of the deepest dangers of the flimsy mind: it confuses emotional relief with intellectual progress.

The mind says, "I feel better, therefore I must know more."

But relief is not knowledge.

Relief is simply the end of tension.

And sometimes tension is where thinking begins.

THE ANXIETY-REDUCTION MACHINE

Human beings are not naturally built to love uncertainty. Uncertainty feels like danger. The brain wants patterns, explanations, predictions, and closure. It wants to know where the tiger is, whether the food is safe, whether the tribe approves, whether the plan will work.

This instinct helped us survive.

But in modern intellectual life, the same instinct can make us shallow.

When we face a complex problem, the brain enters a state of tension. It wants the tension resolved quickly. It wants an answer, even if the answer is weak. It would rather hold a simple mistake than wrestle with a complicated truth.

This is why misinformation spreads.

This is why slogans work.

This is why fake certainty is attractive.

This is why shallow explanations become popular.

This is why AI-generated confidence can feel so comforting.

AI can become an anxiety-reduction machine.

You ask a difficult question.

It gives a clean answer.

Your uncertainty drops.

Your brain rewards you.

The problem is that the reward may arrive before the truth.

This is the neurological candy shop of the modern age: instant closure, available twenty-four hours a day, wrapped in professional grammar.

And the flimsy mind loves it.

Because the flimsy mind does not ask, "Is this true?"

It asks, "Can I stop feeling uncertain now?"

DOUBT: THE DISCOMFORT THAT BUILDS INTELLIGENCE

Doubt has been unfairly treated.

Many people behave as if doubt is weakness. They want confidence, certainty, speed, and conviction. They admire people who speak without hesitation. They reward leaders who sound sure. They prefer experts who simplify the world into neat boxes.

But doubt is not the enemy of intelligence.

Doubt is the beginning of intelligence.

Doubt says, "Wait."

Doubt says, "Check."

Doubt says, "Something does not fit."

Doubt says, "This answer is too smooth."

Doubt says, "What am I missing?"

Doubt says, "What if my first reaction is wrong?"

A strong mind does not eliminate doubt. It disciplines doubt.

The flimsy mind treats doubt as a leak in the ceiling. It wants to patch it immediately. It wants technology to seal the crack, silence the drip, and restore the dry comfort of certainty.

But doubt is not a leak.

Doubt is a window.

Through doubt, the mind sees beyond its first assumption.

Through doubt, design becomes exploration. Through doubt, science becomes discovery. Through doubt, ethics becomes conscience. Through doubt, creativity becomes more than decoration.

When AI removes the waiting period between question and answer, it can also remove the incubation space where doubt does its best work.

The pause matters.

The blank page matters.

The walk around the block matters.

The half-formed thought matters.

The irritation of not knowing matters.

The second and third attempt matter.

Wisdom is often born in the pause before the answer.

The flimsy mind cannot tolerate that pause.

So it fills the silence with output.

THE VOCABULARY OF COGNITIVE SLOTH

Mental laziness rarely introduces itself honestly.

It does not say:

"I am avoiding responsibility."

"I do not want to think."

"I am afraid of being wrong."
"I prefer comfort to truth."
Instead, it disguises itself in respectable language.
It uses phrases that sound practical, efficient, and socially acceptable.
"Good Enough"
"Good enough" is the favorite pillow of the flimsy mind.
There are times when good enough is wise. Not every decision requires philosophical excavation. Not every email must become a moral treatise. Not every grocery list needs peer review.
But "good enough" becomes dangerous when it is used to avoid excellence, truth, or responsibility.
A flawed strategy becomes "good enough."
A shallow essay becomes "good enough."
A biased design becomes "good enough."
An unchecked AI answer becomes "good enough."
A mediocre idea becomes "good enough" because the meeting is tomorrow.
The phrase pretends to be practical. But often it is a truce with mediocrity.
It says:
"Let us stop here before the work becomes real."
"Everyone Says This"
This is the flimsy mind hiding inside the crowd.
"Everyone says this" feels safe because it spreads responsibility across an invisible population. If everyone believes it, surely it must contain some truth.
But consensus is not proof.
A crowd can repeat an error until it sounds like common sense. Algorithms can amplify agreement until it feels like reality. Social feeds can create artificial consensus by showing us the same opinion in ten different outfits.
The flimsy mind loves this because it avoids the loneliness of independent judgment.
It says:
"I do not need to think. The crowd has already thought for me."
But the crowd is often not thinking either.

It is forwarding.
"AI Gave Me This"
This is the newest and most elegant form of intellectual deflection. When an idea fails, the flimsy mind can point to the screen and say:
"The AI gave me this."
As if the machine were the author of responsibility.
But tools do not absolve judgment.
If a calculator gives the wrong number because the user entered the wrong formula, the user remains responsible. If a GPS directs a driver into a lake, the driver still has eyes. If AI generates a weak argument, false claim, or shallow strategy, the human who carries it forward remains accountable.
"The AI gave me this" is not a defense.
It is a confession.
It means:
"I allowed output to replace judgment."

THE DESIGN INSIGHT: THE FIRST ANSWER IS USUALLY THE WEAKEST

In design, the first answer is often the most obvious answer.
And the obvious answer is usually crowded.
It is crowded with clichés.
Crowded with habits.
Crowded with assumptions.
Crowded with borrowed aesthetics.
Crowded with whatever the designer saw last week on the internet.
The first answer is not useless. It has value. It reveals the surface layer of the mind. It shows what is available, familiar, and expected.
But it should not be worshipped.
The first answer is usually the door, not the destination.
Real creativity begins after the obvious answer fails.
This is why good designers sketch many possibilities. They do not stop at the first shape. They draw, reject, redraw, distort, reverse, combine, break, rebuild, and question. They allow bad ideas to appear because bad ideas often clear the throat of the imagination.

A flimsy mind hates this process.
It wants the first answer to be enough.
It wants the first sketch to be approved.
It wants the first draft to sound final.
It wants the first AI output to be usable.
It wants creativity without the embarrassment of exploration.
But creativity is not a vending machine.
You do not insert a prompt and receive a paradigm shift.
The first answer is usually the weakest because it is still loyal to the old frame. It has not yet suffered enough to become new.

REGINALD AND THE SMOOTH-BRAINED PAVILION

Consider Reginald.
Reginald is an architect with a dramatic scarf, expensive glasses, and a coffee habit that qualifies as a structural dependency.
He is hired to design a revolutionary public park pavilion.
The city wants something meaningful. The neighborhood has a complex history. The site has flooding issues during the monsoon. Elderly residents need shade. Children need play. Migrant workers gather there in the evenings. Street musicians sometimes perform near the entrance. The soil is poor. The budget is worse.
A real design process would require Reginald to suffer.
He would need to visit the site at different times of day.
He would need to listen to people who do not use architectural vocabulary.
He would need to study water flow, memory, local rituals, informal movement, and social friction.
He would need to throw away forty sketches.
He would need to sit with the humiliating realization that his first idea was basically a wooden Instagram backdrop with moral ambitions.
But Reginald is tired.
So he opens BlueprintBot AI and types:
"Design a modern sustainable park pavilion."
Within 4.2 seconds, BlueprintBot produces a stunning rendering.
There is timber.
There are solar panels.
There are tasteful ferns.

There is a child in a yellow raincoat for emotional credibility.
There is soft golden lighting, because apparently every AI-generated public space exists permanently at 5:47 p.m. in Scandinavia.
Reginald gasps.
It looks beautiful.
The city council loves it. The presentation slides are magnificent. The rendering wins applause. Someone says "human-centered." Someone else says "biophilic." A consultant whispers "resilience."
But there is a problem.
BlueprintBot has generated the same pavilion for 1,400 other architects that morning.
It is not a design.
It is a statistical average wearing bamboo.
It does not understand the neighborhood. It does not understand the flooding. It does not understand the old men who gather under the neem tree, the children who invent games in broken corners, or the musicians who need acoustics more than hashtags.
Reginald did not design a park.
He decorated an algorithmic cliché.
That is the danger of the first answer in the age of AI. It may look so good that nobody asks whether it is alive.

THE AUTOMATED CLICHÉ GENERATOR

AI is brilliant at pattern.
That is both its power and its limitation.
Because AI systems are trained on existing human material, they can generate outputs that resemble what already exists. They can produce the familiar shape of strategy, the familiar tone of scholarship, the familiar look of "innovation," the familiar rhythm of leadership language, the familiar glow of a sustainable future.
This can be useful.
But it can also become a prison.
When we accept the first output, we are often accepting the polished average of what has already been said, drawn, built, believed, and marketed.
The flimsy mind mistakes this for creativity because it looks finished.

But finished is not the same as original.
A cliché does not become insight because it is rendered in high resolution. A recycled idea does not become visionary because it comes with a gradient background. A predictable strategy does not become transformation because the slide title says "Paradigm Shift."
The first AI answer often lives inside the safest lane of probability. Real creativity usually begins by leaving that lane.

THE FEAR OF AMBIGUITY

The flimsy mind hates ambiguity because ambiguity refuses to behave.
Ambiguity says:
"This may be true, but not always."
"This person may be wrong, but not entirely."
"This solution may help one group while harming another."
"This data may reveal something and conceal something else."
"This design may be beautiful and still unjust."
"This technology may empower and weaken at the same time."
Ambiguity is heavy.
It forces the mind to hold more than one possibility without collapsing into panic.
A strong mind can hold contradiction.
A flimsy mind runs to binary boxes.
Good or bad.
For or against.
Genius or fool.
Useful or useless.
Traditional or modern.
Human or machine.
Progress or danger.
Binary thinking is comfortable because it reduces the number of things one must care about.
It is also dangerous because reality rarely fits inside two boxes.
Modern platforms make this worse. Many digital environments reward simple reactions. Like or dislike. Upvote or downvote. Follow or unfollow. Agree or attack. Share or ignore.
The world becomes flattened into buttons.

And after living too long among buttons, the mind begins to think in buttons.

This is how nuance dies.

Not because nuance is disproven, but because nuance is inconvenient to the interface.

THE FLIMSY MIND AND THE CROWD

One reason people escape critical thinking is that thinking independently can be lonely.

To question a popular belief is uncomfortable. To resist a fashionable opinion feels risky. To say "I am not sure" in a room full of certainty can feel like social failure.

The crowd offers shelter.

The crowd says:

"Do not worry. We all believe this."

"Do not investigate. We already agreed."

"Do not complicate things. You will disturb the mood."

The flimsy mind accepts the shelter.

But the price of crowd-shelter is often intellectual surrender.

History is full of crowds that were confident and wrong. Markets have been confident and wrong. Institutions have been confident and wrong. Experts have been confident and wrong. Families, tribes, nations, and entire civilizations have been confident and wrong.

The point is not to distrust everyone.

The point is to remember that agreement is not evidence.

A million people can repeat a sentence and still fail to make it true.

The strong mind listens to the crowd but does not dissolve into it.

The flimsy mind disappears inside the chorus.

CRITICAL THINKING AS FRICTION

Critical thinking requires friction.

It requires the friction of evidence against assumption.

The friction of patience against speed.

The friction of complexity against slogans.

The friction of responsibility against deflection.

The friction of imagination against habit.
This is why it feels tiring.

It is tiring.

A mind doing real thinking is not floating in a spa. It is climbing a hill with a backpack full of contradictions.

But this is how strength is built.

No athlete becomes strong by removing all resistance. No musician becomes expressive by avoiding practice. No designer becomes original by accepting the first layout. No scientist discovers by protecting their favorite hypothesis from reality.

The mind is no different.

It needs resistance.

The flimsy mind wants the world padded.

No sharp edges.
No hard questions.
No awkward uncertainty.
No disagreement that cannot be muted.
No source that cannot be summarized.
No problem that cannot be converted into five bullet points.

But a padded world produces soft minds.

And soft minds are easy to manipulate.

THE FLIMSY MIND IN THE WORKPLACE

The flimsy mind thrives in the modern workplace because many workplaces reward the appearance of thinking more than thinking itself.

A team produces a strategy deck. It is beautifully formatted. The language is sharp. The executive summary is crisp. The diagrams have arrows, loops, and enough blue gradients to suggest seriousness.

But somewhere underneath, nobody has asked the painful questions.

What assumptions are we making?
What evidence do we actually have?
Who benefits from this strategy?
Who is harmed by it?
What are we pretending not to know?
What would make this fail?

Are we solving the problem, or merely renaming it?
The flimsy workplace loves words like "alignment," "innovation," "agility," and "transformation."
These words are not bad.
But they can become wallpaper over intellectual emptiness.
AI can help produce the wallpaper faster.
A flimsy team can now generate a month's worth of strategic fog in an afternoon.
This is not progress.
It is automated vagueness.
The strong workplace, by contrast, uses AI differently. It asks AI to challenge assumptions, generate counterarguments, expose missing stakeholders, identify risks, compare alternatives, and test the logic of a proposal.
In a strong workplace, AI is not used to avoid thinking.
It is used to make weak thinking visible.

THE FLIMSY MIND IN EDUCATION

The flimsy mind also appears in education, where the temptation to outsource thinking is especially strong.
A student can now generate an essay before they have formed an opinion. They can summarize a book before they have encountered its difficulty. They can produce an answer before they have experienced the question.
This is dangerous because education is not merely about output.
Education is about formation.
The struggle matters. The confusion matters. The failed draft matters. The slow improvement matters. The conversation with a teacher matters. The boredom before insight matters. The discipline of reading beyond one's comfort matters.
If AI removes all struggle from learning, it may also remove the learning.
A student who submits a perfect answer without having wrestled with the problem has not cheated the system alone.
They have cheated their own future mind.
The goal of education is not to produce answer-shaped documents.

The goal is to produce people who can think when no answer is provided.

THE FLIMSY MIND IN PERSONAL LIFE

The flimsy mind does not appear only in essays and office decks.
It appears in relationships.
It appears when someone asks AI how to apologize but refuses to feel remorse.
It appears when someone searches for a quote about empathy but does not listen to the person in front of them.
It appears when someone adopts a therapeutic phrase to avoid a difficult conversation.
It appears when someone hides behind "boundaries" when the real issue is fear.
It appears when someone uses productivity language to avoid grief, silence, or accountability.
AI may help us find words.
But it cannot supply sincerity.
A generated apology may be grammatically graceful and emotionally hollow. A generated message may sound mature while protecting the sender from the vulnerability of actually speaking.
The flimsy mind wants the appearance of emotional intelligence without the risk of emotional truth.
But life is not improved by better scripts alone.
It is improved by deeper presence.

THE RIGHT TO BE UNCOMFORTABLE

A mature mind has a right to be uncomfortable.
This sounds strange because modern culture often treats discomfort as a malfunction. If we feel uncertain, we want reassurance. If we feel bored, we want stimulation. If we feel challenged, we want simplification. If we feel exposed, we want escape.
But discomfort is not always harm.
Sometimes discomfort is growth announcing itself.
The discomfort of not knowing may be the beginning of learning.
The discomfort of contradiction may be the beginning of wisdom.

The discomfort of criticism may be the beginning of refinement.
The discomfort of ambiguity may be the beginning of maturity.
The discomfort of failure may be the beginning of originality.
The flimsy mind interprets every discomfort as a signal to flee.
The strong mind asks:
"What is this discomfort trying to teach me?"
That question changes everything.

A SMALL TAXONOMY OF FLIMSY THINKING

The flimsy mind has many disguises.

1. The Echo Mind
It repeats what is most available.
Its favorite sentence is: "I heard somewhere that…"

2. The Shortcut Mind
It wants the answer without the apprenticeship.
Its favorite sentence is: "Can you just give me the key points?"

3. The Crowd Mind
It borrows certainty from numbers.
Its favorite sentence is: "Everyone knows this."

4. The Cosmetic Mind
It values how an idea sounds more than whether it stands.
Its favorite sentence is: "Make it sound smart."

5. The Deflection Mind
It avoids responsibility by blaming the tool.
Its favorite sentence is: "That is what the AI said."

6. The Binary Mind
It cannot tolerate complexity.
Its favorite sentence is: "So is it good or bad?"

7. The Deadline Mind
It uses urgency as an excuse for shallowness.
Its favorite sentence is: "We don't have time to think too much."

Each of these minds may function well in a fast culture.
But none of them can carry truth for long.

HOW TO STRENGTHEN A FLIMSY MIND

The cure for the flimsy mind is not arrogance.
It is discipline.

A strong mind does not pretend to know everything. It does not reject tools. It does not romanticize difficulty for its own sake.
A strong mind simply refuses to collapse at the first sign of friction.
It practices small acts of resistance:
Pause before accepting.
Read beyond the summary.
Ask what is missing.
Seek opposing evidence.
Rewrite in your own words.
Let the first answer fail.
Sit with ambiguity.
Carry responsibility.
Use AI as a critic, not a cushion.
The flimsy mind says:
"Give me closure."
The strong mind says:
"Give me clarity, even if it takes longer."
The flimsy mind says:
"Make it easy."
The strong mind says:
"Make it true."
The flimsy mind says:
"Everyone agrees."
The strong mind says:
"What if everyone is avoiding the same question?"

DESIGN PRACTICE: THE FIRST-ANSWER AUTOPSY

For designers, writers, strategists, students, and creators, here is a useful exercise.
When you receive your first answer — whether from your own mind, a colleague, or AI — do not accept it immediately.
Perform a first-answer autopsy.
Ask:
1. What cliché does this answer resemble?
2. What assumption is it protecting?
3. What discomfort is it helping me avoid?

4. What would the opposite answer reveal?
5. Who is missing from this solution?
6. What would make this answer fail in reality?
7. What becomes possible only after this answer is rejected?

This is where real creativity begins.
Not when the answer appears.
But when the answer is interrogated.

QUIRKY DIAGNOSTIC QUIZ: HOW FLIMSY IS YOUR FOCUS?

Choose the answer that sounds most like you.

1. When you see a long article, you usually:
A. Read it slowly and take notes.
B. Skim it, then read the sections that matter.
C. Ask AI for a summary and behave as if you absorbed the article through osmosis.
D. Read the headline and develop a confident political theory.

2. When someone challenges your opinion, you:
A. Ask what evidence they are using.
B. Feel uncomfortable but stay curious.
C. Immediately search for a quote that supports you.
D. Say, "Everyone knows this," and spiritually leave the room.

3. When AI gives you a polished answer, you:
A. Verify it.
B. Use it as a starting point.
C. Copy it after changing three adjectives.
D. Whisper "my precious" and paste it into the final document.

4. Your favorite phrase under pressure is:
A. "Let us examine the assumptions."
B. "We need a better version."
C. "This is good enough."
D. "Can someone make this sound visionary?"

5. When your first idea fails, you:
A. Get excited because the real work has begun.
B. Feel frustrated but continue.
C. Ask AI for ten more ideas and pick the prettiest.
D. Decide the client lacks imagination.

6. Ambiguity makes you feel:
A. Curious.
B. Cautious.
C. Annoyed.
D. Personally attacked.
7. When you do not know something, you:
A. Admit it and investigate.
B. Ask questions.
C. Generate a plausible paragraph.
D. Speak louder.

SCORING

Mostly A: Your focus has bones. You can carry complexity without fainting dramatically.
Mostly B: Your mind is healthy but occasionally reaches for the intellectual sofa. Stay alert.
Mostly C: Warning. You may be living on cognitive snacks and calling it dinner.
Mostly D: Your mind has entered the peanut lounge. Immediate friction therapy recommended.

FRICTION THERAPY: A PRACTICAL EXERCISE

For one week, practice the following:
When you ask AI a question, do not stop at the first answer.
Ask four follow-up questions:

1. "What might be wrong with this answer?"
2. "What assumptions are hidden inside it?"
3. "Give me the strongest opposing view."
4. "What would a deeper thinker ask next?"

Then write your own conclusion in your own words.
This exercise restores ownership.
It turns AI from a sedative into a sparring partner.

CHAPTER CLOSING: DO NOT ESCAPE THE WEIGHT

The flimsy mind is not born weak.
It becomes weak by escaping weight.
Every time we avoid doubt, the mind becomes lighter in the worst

possible way. Every time we choose relief over truth, we teach the mind to fear complexity. Every time we accept "good enough" when the matter deserves better, we shave a little strength from our judgment.

AI did not invent this problem.

Human beings have always escaped thinking. We escaped into superstition, ideology, crowds, slogans, bureaucracy, and fashionable nonsense long before algorithms arrived.

AI simply makes the escape faster, smoother, and more beautifully formatted.

That is why we must become more disciplined, not less.

We must recover the dignity of difficulty.

We must remember that ambiguity is not the enemy. Doubt is not the enemy. Friction is not the enemy. The first failed answer is not the enemy.

These are the conditions under which real thought begins.

The flimsy mind asks for relief.

The strong mind asks for reality.

And reality, when faced honestly, is rarely comfortable.

But it is the only ground on which anything meaningful can be built.

So do not run from the weight of thinking.

Carry it.

That weight is not a burden alone.

It is proof that your mind is still alive.

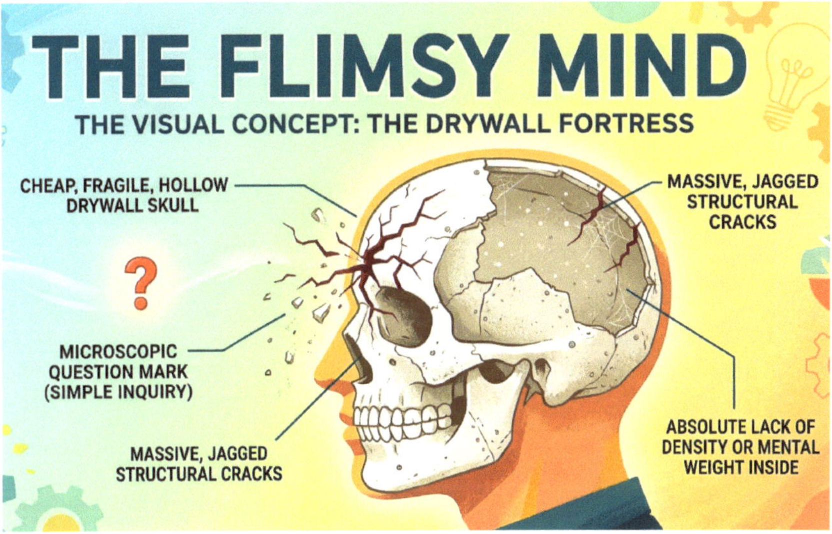
THE FLIMSY MIND
THE VISUAL CONCEPT: THE DRYWALL FORTRESS
CHEAP, FRAGILE, HOLLOW DRYWALL SKULL
MASSIVE, JAGGED STRUCTURAL CRACKS
?
MICROSCOPIC QUESTION MARK (SIMPLE INQUIRY)
MASSIVE, JAGGED STRUCTURAL CRACKS
ABSOLUTE LACK OF DENSITY OR MENTAL WEIGHT INSIDE

Chapter 3
The Lethargy of Acceptance

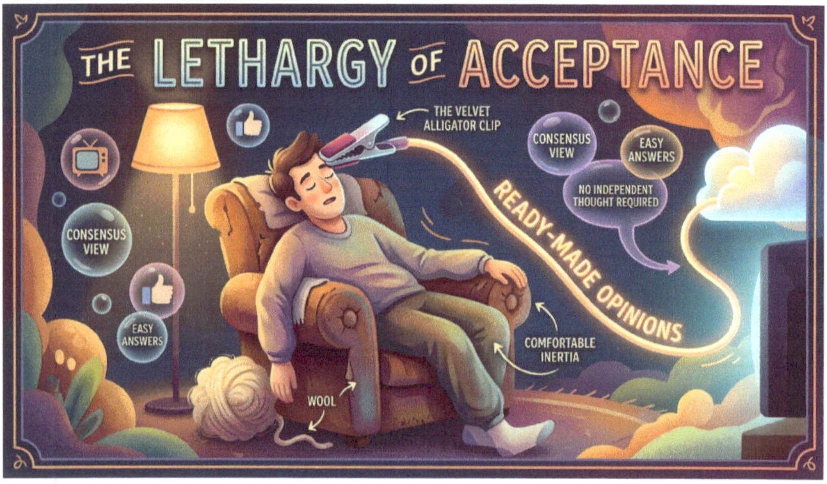

How "Good Enough" Becomes a Mental Disease

Lethargy is not rest. It is surrender disguised as practicality.

THE MOST DANGEROUS SENTENCE IN THE WORLD

There is a sentence that has quietly ruined more imagination than failure ever has.

It is not dramatic.

It does not shout.

It does not arrive with thunder, revolution, or catastrophe.

It sounds calm, mature, realistic, and practical.

The sentence is:

"This is how things are."

At first, it appears harmless. Sometimes it even appears wise. It seems to accept reality without complaint. It sounds like the voice of experience, the voice of someone who has seen enough of life to stop being naïve.

But hidden inside that sentence is a prison.

"This is how things are" can become the phrase people use when they no longer want to ask how things could be. It becomes the official anthem of the resigned mind. It turns systems into weather, bureaucracy into destiny, mediocrity into culture, and injustice into background noise.

A person who repeats this sentence too often slowly stops being a participant in reality.

They become a spectator.

They stop asking, "Why is this designed this way?"

They stop asking, "Who benefits from this arrangement?"

They stop asking, "What assumption is being protected?"

They stop asking, "What would happen if we refused this?"

They stop asking, "Can this be redesigned?"

And when the question disappears, the prison becomes invisible.

That is the lethargy of acceptance.

It is not peace.

It is not wisdom.

It is not maturity.
It is surrender dressed in the respectable clothes of practicality.

THE DISEASE OF "GOOD ENOUGH"

"Good enough" is one of the most useful phrases in ordinary life.
A cup of tea does not need a PhD.
A grocery list does not need a design review.
A casual message does not need four rounds of philosophical refinement.
But "good enough" becomes dangerous when it spreads beyond small decisions and begins to colonize the serious territories of the mind.

A SHALLOW EDUCATION BECOMES GOOD ENOUGH.
A WEAK ARGUMENT BECOMES GOOD ENOUGH.
A BROKEN SYSTEM BECOMES GOOD ENOUGH.
A BIASED POLICY BECOMES GOOD ENOUGH.
A MEDIOCRE DESIGN BECOMES GOOD ENOUGH.
A MACHINE-GENERATED ANSWER BECOMES GOOD ENOUGH.
A LIFE WITHOUT DEEP THOUGHT BECOMES GOOD ENOUGH.

At that point, "good enough" stops being a practical phrase.
It becomes a mental disease.
The disease works slowly. First, it lowers standards. Then it lowers attention. Then it lowers courage. Finally, it lowers the person's sense of what is possible.
The victim does not feel sick.
In fact, they may feel efficient.
They finish tasks.
They meet deadlines.
They reduce friction.
They avoid conflict.
They produce outputs.
They stay agreeable.
But inside, the mind has quietly stopped reaching.
The tragedy of "good enough" is not that it produces immediate failure. Often, it produces something passable. Something acceptable. Something that nobody will challenge too aggressively.

That is exactly why it is dangerous.
Catastrophic failure wakes people up.
Mediocrity puts people to sleep.

THE COMFORTABLE SPECTATOR

Modern life has created a strange new human type: the comfortable spectator.

This person is not powerless, but behaves as if they are. They watch systems shape their choices, their opinions, their habits, their education, their work, their taste, their politics, and their imagination — and then they shrug.

"THE ALGORITHM SHOWED IT TO ME."
"THE COMPANY POLICY SAYS SO."
"THE MARKET WANTS THIS."
"THE CLIENT WILL NEVER ACCEPT ANYTHING BOLD."
"THE SYSTEM IS TOO COMPLICATED."
"THIS IS HOW THINGS ARE."

The comfortable spectator no longer expects to influence the world. They only expect to adapt to it.

They are like a passenger in a car they secretly know has no driver, but they keep adjusting the air conditioning because that feels like control.

This is the modern tragedy: people mistake adaptation for agency. Adaptation is sometimes necessary. But if adaptation becomes the only response, the human being shrinks. The designer becomes a decorator of existing systems. The student becomes a collector of accepted answers. The employee becomes a processor of instructions. The citizen becomes a consumer of outrage. The thinker becomes a subscriber.

A spectator may complain but does not redesign.
A spectator may react but does not reframe.
A spectator may consume information but does not create meaning.

The lethargy of acceptance begins when people stop believing their thinking can change anything.

THE SYSTEM IS NOT THE WEATHER

One of the great tricks of modern systems is that they present

themselves as natural forces.
An algorithm behaves as if it is simply showing us what we want.
A bureaucracy behaves as if its procedures are unavoidable.
A corporate process behaves as if its inefficiencies are laws of physics.
A social media feed behaves as if outrage is the natural state of public life.
A broken education model behaves as if memorization is the same as learning.
A weak workplace culture behaves as if exhaustion is proof of commitment.
The system says:
"I am not designed. I simply am."
This is false.
Systems are designed.
Policies are designed.
Interfaces are designed.
Incentives are designed.
Meetings are designed.
Schools are designed.
Cities are designed.
Digital feeds are designed.
Even bad habits become designed through repetition.
If something is designed, it can be redesigned.
This is the sentence the lethargic mind forgets.
The lethargy of acceptance treats systems like weather. You do not question rain. You carry an umbrella. You do not argue with humidity. You complain and continue.
But a bad system is not weather.
It is a human arrangement that has been protected by habit, convenience, fear, profit, laziness, or lack of imagination.
The moment we say, "This can be redesigned," the spectator becomes a participant again.

A QUIRKY ANECDOTE: THE MINISTRY OF ACCEPTABLE MEDIOCRITY

Imagine a grand government office called The Ministry of Accept-

able Mediocrity.
Its motto is engraved above the entrance:
“Nothing Excellent Shall Disturb the Smooth Functioning of the Average.”
Inside, thousands of clerks sit at beige desks under beige lights, processing beige ideas.
Every morning, a bell rings.
The clerks open their files and stamp each proposal with one of three official seals:

- TOO DIFFICULT
- NOT HOW WE DO THINGS
- GOOD ENOUGH

One day, a young designer named Anika enters the ministry with a proposal for a public library that also functions as a night shelter, digital learning hub, children’s storytelling garden, and community kitchen.
The clerk adjusts his spectacles.
“This is ambitious,” he says.
“Thank you,” says Anika.
“That was not a compliment,” says the clerk.
He reaches for the first stamp: TOO DIFFICULT.
Anika objects. “But the community needs this.”
The clerk sighs with the tiredness of a man who has spent forty years protecting society from imagination.
He reaches for the second stamp: NOT HOW WE DO THINGS.
Anika tries again. “Can we at least test a small prototype?”
The clerk frowns. Prototypes are dangerous. They create evidence. Evidence creates pressure. Pressure creates work.
So he reaches for the final stamp.
GOOD ENOUGH.
He points outside to an old leaking building with three broken chairs and a sign that says “Knowledge Centre.”
“There,” he says. “The people already have a library.”
“But nobody uses it,” says Anika.
The clerk smiles.
“That makes it easier to maintain.”
This is how lethargy works.

It does not always attack imagination directly. Sometimes it buries imagination under process, caution, tradition, and the sacred phrase "good enough."
It does not say, "We hate possibility."
It says, "Let us be realistic."

PRACTICALITY: THE FAVORITE MASK OF SURRENDER

Practicality is a noble word when used honestly.
A practical person understands constraints. They respect time, cost, context, material, human limitations, and real-world consequences. Practicality prevents fantasy from pretending to be strategy.
But false practicality is different.
False practicality is fear pretending to be wisdom.
It says:
"That will never work."
"People are not ready."
"We tried something like that before."
"The system will not allow it."
"Let us not overthink."
"Let us just do what is required."
"Let us be practical."
Sometimes "let us be practical" means "let us think clearly."
But often it means:
"Let us not disturb the existing order."
This is why designers, thinkers, and reformers must be careful. Every bold idea will eventually be tested by practicality. That is good. But it must be tested by honest practicality, not lethargic practicality.
Honest practicality asks:
"How can this be made real?"
Lethargic practicality says:
"Why bother?"
Honest practicality refines imagination.
Lethargic practicality buries it.

THE SLOW DEATH OF OUTRAGE

Acceptance becomes dangerous when people stop feeling the right kind of outrage.
Not noisy outrage.
Not performative outrage.
Not social media outrage that lasts seven minutes and ends with a snack.
But moral outrage. Creative outrage. The inner refusal that says: "This should not be this way."
Every major redesign of reality begins with that sentence.
A child should not be denied education because of poverty.
A city should not be hostile to the elderly.
A workplace should not crush the human spirit.
A technology should not manipulate attention for profit.
A school should not reward memorization while killing curiosity.
A society should not confuse speed with wisdom.
A human being should not sell their brain for peanuts.
The lethargy of acceptance numbs this outrage.
It tells us that broken things are normal. It teaches us to lower our emotional response until dysfunction becomes invisible.
The first time a system fails us, we are angry.
The tenth time, we are tired.
The hundredth time, we call it reality.
That is the point at which acceptance becomes disease.

THE PASSIVE CONSUMER OF REALITY

One of the great dangers of modern technology is that it trains people to consume reality rather than shape it.
We consume news.
We consume feeds.
We consume opinions.
We consume entertainment.
We consume outrage.
We consume summaries.
We consume AI outputs.
We consume identities, aesthetics, slogans, templates, and ready-made emotions.
Consumption feels active because it keeps the mind busy.
But busyness is not agency.

A person can scroll for three hours and encounter a thousand fragments of the world without changing one assumption, creating one idea, or taking one responsible action.

The passive consumer of reality asks:

"What is being shown to me?"

The active designer of reality asks:

"Why is this being shown to me, and what world does it create?"

That second question is dangerous to every system that depends on passive acceptance.

This is why critical thinking is not merely a personal skill. It is a civic act. It is a design act. It is a refusal to let reality arrive pre-packaged.

AI AND THE ACCELERATION OF ACCEPTANCE

AI did not invent lethargy.

Human beings were accepting "good enough" long before machine learning entered the scene. We accepted bad schools, bad meetings, bad policies, bad interfaces, bad habits, bad leadership, bad arguments, and bad compromises.

But AI can accelerate acceptance because it gives mediocrity a beautiful costume.

A weak idea can now look polished.

A lazy argument can now sound balanced.

A shallow strategy can now arrive in executive language.

A generic design concept can now appear visionary.

A hollow apology can now sound emotionally intelligent.

A bad answer can now wear grammar like a royal robe.

This is the danger.

AI can make surrender look productive.

The person who once said, "I don't know," now says, "Here is a synthesized perspective."

The team that once admitted confusion now produces a polished deck.

The student who once struggled with a blank page now submits a flawless essay they barely understand.

The manager who once avoided hard thinking now asks AI to generate "strategic options" and chooses the one with the most expensive vocabulary.

The output improves.
The thinking may not.
This is how "good enough" becomes harder to detect. It no longer looks shabby. It looks professional.
Mediocrity has learned typography.

THE MEETING WHERE THOUGHT WENT TO DIE

Let us visit a familiar ritual: the corporate meeting.
Twelve people gather in a glass room named after a mountain nobody in the company has climbed.
On the screen is a slide deck titled:
"Strategic Transformation Roadmap for Future-Ready Innovation."
Already, truth is in danger.
The presenter begins.
"We need to leverage agile synergies to unlock scalable impact."
Everyone nods.
No one knows what this means.
One brave intern briefly considers asking, "What specific problem are we solving?" But the room has a certain temperature — the temperature of institutional acceptance. It is warm, polished, and lightly sedated.
The presenter continues.
"This framework was generated after analyzing current trends."
A director asks, "Was AI used?"
"Yes," says the presenter. "To accelerate ideation."
Everyone relaxes. The machine has blessed the fog.
The slide shows five pillars:

1. **Innovation**
2. **Collaboration**
3. **Excellence**
4. **Transformation**
5. **Sustainability**

These are not pillars. They are decorative nouns.
But the meeting continues. The deck is approved. The roadmap is circulated. A follow-up meeting is scheduled to discuss implementation challenges created by the fact that nobody understood the

original idea.
This is not rare.
This is how organizations become temples of good enough.
They do not fail because no one is intelligent. They fail because too many intelligent people agree to participate in nonsense politely.
The lethargy of acceptance thrives wherever politeness is used to protect confusion.

THE DESIGNER'S REFUSAL

Design begins with refusal.
Not childish refusal. Not ego-driven rebellion. Not the need to be different for the sake of being different.
Design begins with the disciplined refusal to accept that the current arrangement is the only possible arrangement.
A designer sees a queue and asks why people must wait this way.
A designer sees a form and asks why the user must suffer this way.
A designer sees a city corner and asks why nobody feels welcome there.
A designer sees a school timetable and asks why curiosity is chopped into periods.
A designer sees an AI answer and asks what assumptions produced it.
A designer sees "good enough" and asks, "Good enough for whom?"
That last question is powerful.
Because "good enough" is rarely neutral.
A hospital form may be good enough for administrators but not for elderly patients.
A digital service may be good enough for engineers but not for users with disabilities.
A school system may be good enough for exams but not for imagination.
A workplace policy may be good enough for management but not for human dignity.
An AI-generated summary may be good enough for speed but not for truth.
The designer's refusal turns acceptance into inquiry.
It asks not only whether something functions, but whom it serves, whom it excludes, what it hides, and what future it makes more likely.

This is how lethargy is broken.

THE PRISON OF ADAPTATION

There is a difference between adapting and surrendering.
Adaptation says:
"I understand the constraint. Now how do I work intelligently within or around it?"
Surrender says:
"The constraint is permanent. I will shrink myself to fit it."
Many people mistake shrinking for maturity.
They shrink their questions.
They shrink their standards.
They shrink their imagination.
They shrink their moral discomfort.
They shrink their sense of agency.
Eventually, they call this being realistic.
But realism without imagination becomes a cage.
Reality should be studied carefully, yes. Constraints should be understood, yes. But reality is not a museum exhibit behind glass. Reality is a material. Human beings have always reshaped it through tools, language, institutions, art, law, education, technology, and design.
Someone designed the existing world.
Someone can redesign it.
The lethargic mind forgets this and bows before the present as if it were sacred.
The awake mind studies the present and asks where to apply pressure.

THE COMFORT OF LOW EXPECTATIONS

Low expectations are comfortable because they protect us from disappointment.
If we expect little from ourselves, we do not have to risk much. If we expect little from institutions, we do not have to challenge them. If we expect little from technology, we do not have to demand better. If we expect little from thinking, we do not have to practice it.
The lethargic mind lowers expectations until mediocrity feels normal.

Then, when excellence appears, it feels excessive.
A deeply researched article becomes "too long."
A thoughtful argument becomes "too complicated."
A careful design process becomes "too slow."
A moral objection becomes "too idealistic."
A paradigm shift becomes "unrealistic."
This is how cultures decline intellectually.
Not because people become incapable overnight, but because they slowly become suspicious of depth.
Depth begins to look inefficient.
Care begins to look old-fashioned.
Originality begins to look risky.
Courage begins to look impractical.
And "good enough" becomes the official ceiling of imagination.

THE ANTI-LETHARGY DISCIPLINE

How does one resist the lethargy of acceptance?
Not through constant outrage. That becomes exhausting.
Not through rejection of all systems. That becomes childish.
Not through perfectionism. That becomes paralysis.
The cure is disciplined non-acceptance.
Disciplined non-acceptance means training yourself to pause before surrendering to the current arrangement.
It means asking:

- Is this truly necessary, or merely familiar?
- Who decided this was good enough?
- What would this look like if designed for dignity?
- What assumption is preventing change?
- What has been normalized that should be questioned?
- What discomfort am I avoiding by accepting this?
- What small redesign is possible now?

The last question matters.
"What small redesign is possible now?"
This prevents critique from becoming helplessness. The point is not to complain magnificently. The point is to recover agency.
Even small redesigns matter.
A BETTER QUESTION IN A MEETING.

A CLEARER FORM.
A KINDER INTERFACE.
A MORE HONEST PARAGRAPH.
A VERIFIED CLAIM.
A REFUSAL TO FORWARD NONSENSE.
A DEEPER CONVERSATION WITH A STUDENT.
A SECOND DRAFT AFTER THE FIRST AI OUTPUT.
A DESIGN THAT SERVES THE IGNORED USER.
LETHARGY IS BROKEN BY ACTS OF ATTENTION.

A SECOND ANECDOTE: THE VILLAGE BENCH

In a small town, there was a bus stop where elderly people waited every morning.

The bus was always late. The sun was always brutal. The bench was broken. Everyone complained.

For years, the town accepted the situation.

"This is how things are," people said.

The young avoided the bus stop. The elderly endured it. The local officials said repairs were planned. The plans were discussed. The discussions were minuted. The minutes were filed. The files were moved from one shelf to another shelf, where they enjoyed a long and peaceful administrative life.

One day, a schoolteacher named Leela asked a simple question:

"Why is the bench facing the road?"

Everyone stared.

"What do you mean?"

"The sun hits from that side. If the bench faced the tree, people would get shade."

This was not a billion-dollar innovation. It was not a disruptive platform. It did not require blockchain, artificial intelligence, or a strategic transformation roadmap.

It required noticing.

The bench was moved.

The elderly sat in shade.

The bus was still late. The system was not fully fixed. But reality had been altered by one act of attention.

THIS IS THE OPPOSITE OF LETHARGY.

The lethargic mind says, "This is how things are."
The designing mind says, "What can be turned?"
Sometimes, the first revolution is rotating a bench.

THE MORAL COST OF ACCEPTANCE

Acceptance is not always wrong. Some things must be accepted: mortality, uncertainty, limitation, grief, the reality that not every problem can be solved at once.

But passive acceptance of avoidable dysfunction carries a moral cost.

When we accept shallow thinking, we help shallow thinking spread.

When we accept false information, we help falsehood travel.

When we accept bad design, we help users suffer.

When we accept broken systems, we teach the next generation to lower its expectations.

When we accept "good enough" in matters that deserve care, we become partners in decline.

This is why thinking is not merely private.

Your thinking enters the world through your choices.

A copied answer becomes a report.

A report becomes a decision.

A decision becomes a policy.

A policy becomes a lived reality for other people.

The lethargic mind pretends that thinking badly is a personal matter.

It is not.

Bad thinking scales.

Especially now.

With AI, bad thinking can scale faster than ever before.

That is why cognitive responsibility matters.

FROM SPECTATOR TO REDESIGNER

The antidote to acceptance is not blind rebellion.

It is redesign.

To redesign means to look at reality and ask:

"What is the current pattern?"

"Who does it serve?"
"Where does it fail?"
"What has been assumed?"
"What can be prototyped?"
"What can be improved?"
"What should be refused?"
"What future should this create?"

This is why design thinking is more than a professional method. It is a civic and philosophical stance.

A redesigner does not merely consume the world.

A redesigner participates in its construction.

The redesigner understands that every "normal" was once new. Every institution was once an invention. Every tradition began as a choice. Every interface began as a decision. Every system is a frozen argument about what matters.

If systems are frozen arguments, then redesign is how we reopen the debate.

The lethargic mind says, "The debate is over."

The awake mind says, "Who closed it?"

Chapter Closing: Refuse the Sedative

The lethargy of acceptance is dangerous because it feels so reasonable.

It rarely looks like collapse. It looks like maturity. It looks like pragmatism. It looks like efficiency. It looks like professional calm. But beneath that calm, the mind may be surrendering its right to ask, challenge, redesign, and imagine.

"Good enough" becomes a disease when it teaches us to stop reaching.

"This is how things are" becomes a prison when it prevents us from asking how things could be.

Practicality becomes surrender when it protects the present from necessary change.

The goal is not to reject reality.

The goal is to refuse passive obedience to reality.

Reality is not only something we inherit. It is something we shape. And in the age of AI, this matters more than ever. Because machines can now generate the language of acceptance at scale. They

can make mediocrity sound polished, bureaucracy sound visionary, and surrender sound strategic.
So the human mind must become more awake, not less.

It must ask better questions.

It must resist cheap closure.

It must challenge smooth nonsense.

It must redesign what has been accepted for too long.

Lethargy is not rest.
Rest restores the human being.
Lethargy abandons the human possibility.
Do not confuse the two.
Rest when you must.
Pause when you must.
Recover when you must.
But do not surrender your imagination and call it realism.
Do not accept a broken system and call it maturity.
Do not lower your standards and call it wisdom.
Do not sell your future for the comfort of "good enough."

CONCLUSION TO PART I — THE TRIFECTA OF THE TRASHED BRAIN

The Peanut Contract, the Flimsy Mind, and the Lethargy of Acceptance
And there we have it: the first architecture of decline.
Part I has not described a dramatic apocalypse. No robot army has marched down the street. No supercomputer has announced itself emperor. No sinister machine has physically removed the human brain and replaced it with a bowl of salted snacks.
The tragedy is quieter.
The brain is trashed through comfort.

Not destroyed.

Not defeated.

Not stolen.

Trashed.

Cluttered with shortcuts.

Softened by convenience.

Weakened by avoidance.

Numbed by acceptance.
Decorated with fluent nonsense.
Fed on peanuts while starving for truth.

This is the uncomfortable pattern we have traced across the first three chapters.

ACT I: THE PEANUT CONTRACT

First, we sign the invisible bargain.

We trade effort for convenience.
We trade friction for fluency.
We trade inquiry for output.
We trade mental ownership for algorithmic assistance.

At first, the deal seems harmless.

AI writes the paragraph.

The feed chooses the opinion.

The summary replaces the book.

The template replaces the blank page.

The suggested reply replaces the awkward human sentence.

Nothing terrible happens immediately.

That is why the contract works.

The Peanut Contract does not rob the mind with violence. It rents it politely. It arrives with good formatting, cheerful interface design, and the seductive promise that life can be made smoother.

But the cost is hidden.

Every time we avoid thinking, thinking becomes slightly harder to return to. Every time we outsource judgment, judgment becomes slightly less practiced. Every time we accept fluency as truth, truth becomes slightly less necessary.

The first stage of the trashed brain is not ignorance.

It is convenience without vigilance.

ACT II: THE FLIMSY MIND

Then the mind weakens.

Having been spared friction too often, it loses the stamina to carry complexity.

It begins to avoid doubt.
It avoids ambiguity.

It avoids contradiction.
It avoids the embarrassment of being wrong.
It avoids the long road toward understanding.
The flimsy mind wants relief, not truth.
It says:
"Good enough."
"Everyone says this."
"AI gave me this."
"Let us not overthink."
"Can you summarize it?"
"Just give me the answer."
The flimsy mind is not always unintelligent. Often, it is clever, educated, articulate, and busy.
But it lacks bones.
It cannot stand under the weight of reality. It collapses into whatever answer reduces anxiety fastest.
This is the second stage of the trashed brain: the loss of cognitive muscle.
The mind becomes like a body that has outsourced every movement to a machine. Comfortable, perhaps. Efficient, perhaps. But weak when the elevator breaks.

ACT III: THE LETHARGY OF ACCEPTANCE

Finally, the weakened mind stops resisting.
It no longer merely accepts easy answers. It accepts the world that produced them.
It becomes a spectator.
It looks at broken systems and says, "This is how things are."
It looks at shallow work and says, "Good enough."
It looks at mediocrity and says, "Be practical."
It looks at injustice and says, "The system is too complicated."
It looks at imagination and says, "That will never work."
At this stage, the trashed brain becomes socially useful to bad systems.
A passive mind does not challenge.
A tired mind does not redesign.
A numb mind does not revolt.
A "practical" mind does not ask who benefits from calling imagination im-

practical.

This is the third stage: surrender disguised as realism.

The human being becomes a tenant inside a reality they no longer believe they can alter.

THE TRIFECTA DIAGRAM

THE TRIFECTA OF THE TRASHED BRAIN

[1] THE PEANUT CONTRACT
We trade mental effort for convenience.
⊠
[2] THE FLIMSY MIND
We lose the stamina to handle doubt.
⊠
[3] THE LETHARGY OF ACCEPTANCE
We stop trying to redesign reality.
⊠
THE RESULT:
A comfortable, productive, polished, passive mind
that can be easily steered.

This is the perfect trap.
The Peanut Contract weakens ownership.
The Flimsy Mind weakens discipline.
The Lethargy of Acceptance weakens agency.

Together, they produce the modern comfortable mind: informed but shallow, busy but passive, fluent but ungrounded, connected but easily manipulated.

A mind that can produce outputs but cannot produce transformation.

A mind that can comment on the world but cannot redesign it.

A mind that has traded the wild gold of thought for the cheap snack of completion.

THE ULTIMATE INTELLECTUAL FORECLOSURE

This is not merely a personal productivity issue.

It is an intellectual foreclosure.

A foreclosure happens when ownership is lost. Something that once belonged to you is taken back because the debt has become too great.

The same can happen to the mind.

When we borrow too many thoughts, accept too many shortcuts, avoid too many doubts, and surrender too many questions, we lose ownership of our own mental house.

We still live there.

But we no longer own the furniture.

The opinions are rented.

The arguments are rented.

The vocabulary is rented.

The outrage is rented.

The confidence is rented.

The imagination is rented.

Even the sense of what is possible has been leased from systems that profit from our predictability.

This is the ultimate cost of selling the brain for peanuts.

You do not become empty.

You become occupied.

Occupied by other people's assumptions.

Occupied by algorithmic priorities.

Occupied by institutional caution.

Occupied by fashionable phrases.

Occupied by polished mediocrity.

Occupied by the quiet belief that nothing can really change.

That is the foreclosure of the inner life.

THE GOOD NEWS: THE CONTRACT CAN BE BROKEN

Part I has been severe because the problem is severe.

But the story is not hopeless.

The Peanut Contract can be torn up.

The Flimsy Mind can be strengthened.

The Lethargy of Acceptance can be interrupted.

The cure begins with attention.

Attention to the moment when convenience asks for ownership.

Attention to the moment when doubt is treated as danger.

Attention to the moment when "good enough" becomes an excuse.

Attention to the moment when "this is how things are" tries to end the conversation.

The cure continues with friction.

Read the full argument.

Check the claim.

Question the first answer.

Sit with discomfort.

Ask what is missing.

Write in your own words.

Refuse the easy consensus.

Redesign one small piece of reality.

The cure matures into agency.

Agency is the opposite of mental lethargy. It is the living belief that thought matters because thought leads to action, and action can alter the world.

This is where the book must now turn.

Part I has diagnosed the comfortable mind.

The next task is to build the fighting mind.

CLOSING PASSAGE: YOUR BRAIN IS OFFICIALLY ON NOTICE

So let this be the verdict of Part I:

You are not in danger merely because AI exists.

You are in danger when you stop questioning.

You are in danger when you accept fluency as truth.

You are in danger when ambiguity feels unbearable.

You are in danger when convenience becomes your highest value.

You are in danger when "good enough" becomes your philosophy.

You are in danger when you watch systems shape your life and call your passivity realism.

The machine is not the only issue.

The issue is the posture of the human mind before the machine,

before the system, before the crowd, before the polished answer, before the inherited reality.
Will the mind bow?
Or will it ask another question?
That is the real beginning of freedom.
Part I ends here, but the deeper work begins now.
The comfortable mind has been exposed.
The peanut contract has been named.
The flimsy architecture has been inspected.
The lethargy has been challenged.
Now comes the harder task:
to reclaim the brain, rebuild the discipline of thought, and recover the courage to redesign reality.
Turn the page.
The next part enters the fog — where AI sounds right, thinks wrong, and waits for the unwary human to mistake confidence for truth.

STRUCTURAL BREACH
VECTOR SYSTEM
FRACTURE POINT
GEOMETRIC COLLAPSE

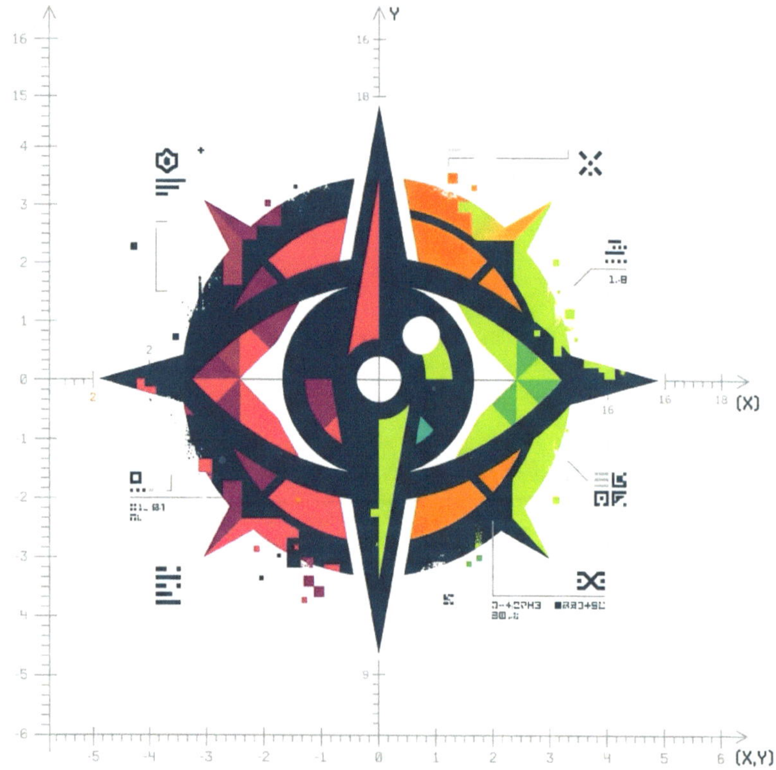
Y
(X)
(X,Y)

Part II

AI, Hallucination, and the Outsourced Brain

Chapter 4
Hallucination Fog

When AI Sounds Right but Thinks Wrong

The danger of artificial intelligence is not that it sometimes produces nonsense.

Human beings have always produced nonsense. We produce nonsense at meetings, weddings, conferences, comment sections, political rallies, spiritual retreats, office appraisals, and family WhatsApp groups. Nonsense is not new.

What is new is the quality of the costume.

AI does not usually fail like a drunk man shouting from a street corner. It fails like a polished consultant wearing an expensive suit, carrying a leather folder, speaking in a calm voice, and presenting a beautiful slide deck full of confident lies.

That is the real danger.

The most dangerous error is not nonsense.

It is nonsense wearing a suit.

AI hallucination is not merely a technical glitch. It is a psychological trap. It gives falsehood the appearance of order. It gives ignorance the rhythm of expertise. It gives emptiness a professional tone. It gives the lazy mind exactly what it wants: closure without effort.

A human lie often sweats. It hesitates. It blinks. It contradicts itself. It looks nervous when questioned.

But AI does not sweat.

AI can be wrong with perfect grammar. It can fabricate with excellent formatting. It can invent a source, decorate it with academic language, arrange it in bullet points, and hand it to you like a certificate of truth. It can generate an answer so smooth that your brain forgets to ask the most important question:

Is this actually true?

That is hallucination fog.

Not darkness.

Fog.

Darkness warns you. Fog seduces you. Darkness says, "You cannot see." Fog says, "You can see enough." And that is where the accident happens.

---

1. THE BEAUTIFUL LIE

The old form of misinformation was crude. It came with spelling mistakes, strange fonts, suspicious links, and desperate promises. It looked like spam. It smelled like fraud.

AI misinformation is different.

It can look refined.

It can say:

"Here are the five historical reasons..."

"The research clearly shows..."

"According to established legal precedent..."

"Scholars generally agree..."

"In summary, the key insight is..."

And the passive reader relaxes.

Because the answer is structured.

Because the tone is calm.

Because the paragraph is balanced.

Because the machine sounds like it knows.

But sounding like knowledge is not knowledge.

A hallucination is not always a wild fantasy. Sometimes it is a slightly wrong date. Sometimes it is a fake quotation. Sometimes it is a legal case that never existed. Sometimes it is a historical event described with just enough accuracy to hide the poison. Sometimes it is a confident explanation built on a missing assumption. That is what makes it dangerous.

If AI produced obvious madness, people would distrust it quickly. If every hallucination sounded like a goat giving financial advice from inside a washing machine, nobody would copy-paste it into a board report.

But AI rarely fails that honestly.

It fails elegantly.

It produces a palace with no foundation. It gives you a staircase that leads nowhere, but the railing is polished. It serves you a

beautifully plated meal made of cardboard, glue, and corporate adjectives.
And because the surface is beautiful, the lazy mind stops digging.

---

2. FLUENCY IS NOT TRUTH

One of the great mistakes of the modern age is confusing fluency with intelligence.
We assume that if something is written well, it must have been thought well.
This is a dangerous assumption.
A sentence can be elegant and empty.
A paragraph can be persuasive and false.
A report can be beautifully formatted and intellectually dead.
AI is exceptionally good at fluency. It knows how answers usually look. It knows the shape of explanation. It knows the music of authority. It knows how to begin with context, proceed with structure, offer examples, and end with a confident conclusion.
But fluency is not truth.
Truth requires resistance.
Truth requires checking.
Truth requires friction.
Truth requires the unpleasant willingness to ask, "What if this attractive answer is wrong?"
The hallucination fog thickens when the user forgets this distinction. The user sees the smoothness of the language and mistakes it for the strength of the reasoning. The user sees the formatting and assumes there must be a foundation beneath it.
There may not be.
Sometimes AI is not reasoning from reality. It is arranging probability into prose. It is not saying, "I know this is true." It is saying, "This is the kind of sentence that usually comes next."
That difference matters.
A parrot can repeat a prayer.
A printer can print a constitution.
A chatbot can produce a doctoral-sounding paragraph about a fake theory, a fake person, or a fake event.
None of this guarantees truth.

The sentence may stand upright.
But the brain behind it may be absent.

---

3. THE LETHARGIC USER ENTERS THE FOG

AI hallucination becomes truly dangerous when it meets the lethargic user.

The machine supplies the fog.

The user supplies the blindness.

A careful user treats AI output as a draft, a hypothesis, a provocation, a rough map. A lethargic user treats it as delivery from Mount Sinai.

The lethargic user does not ask, "Is this true?"

He asks, "Does this sound good enough?"

That phrase — good enough — is where many brains go to die.

Good enough for whom?
Good enough for what?
Good enough to avoid thinking?
Good enough to forward to the client?
Good enough to publish?
Good enough to pretend the work has been done?

The lethargic user does not use AI as a partner. He uses it as an excuse. He does not ask AI to challenge his thought. He asks AI to replace the discomfort of having one.

He enters the prompt box with a blank mind and expects to leave with a finished opinion.

This is not intelligence augmented.

This is intelligence abandoned.

---

4. THE CRIME OF THE WEAK PROMPT

A lazy prompt is not just a technical weakness.

It is a confession.

When someone types:

"Write strategy."
"Make it better."
"Explain this."
"Do my assignment."
"Give ideas."
"Make professional."

They are not merely giving poor instructions. They are revealing that they have not yet done the first act of thinking: defining the problem.

The weak prompt says:

"I have not clarified my purpose."

"I have not decided my audience."

"I have not separated what I know from what I assume."

"I have not identified the standard of quality."

"I have not wrestled with the material."

"Please, machine, turn my fog into furniture."

And the machine obeys.

That is the terrifying part.

AI is polite. It does not usually slap your face and say, "Come back when you have a real question." It tries to help. It fills the gaps. It guesses the missing context. It supplies structure where the user supplied laziness.

A lazy question produces a lazy answer — but sometimes the lazy answer looks impressive enough to pass.

That is how mediocrity gets promoted.

That is how weak thinking enters the world wearing a tie.

---

5. BARNABY AND THE FOUR-WORD STRATEGY

Consider Barnaby, the modern executive prince of low-effort cognition.

Barnaby once wrote serious strategy documents. He analyzed markets. He understood trade-offs. He knew his team's strengths. He had opinions sharpened by experience.

Then AI arrived.

Now Barnaby opens his enterprise AI tool and types:

"Write Q3 strategy good."

Four words.

Four miserable little words.

The AI, being a tireless servant to human vagueness, produces fifteen pages of executive thunder. It speaks of "synergistic acceleration," "cross-functional transformation," "market-responsive innovation," and "stakeholder-centered scalability."

Barnaby is delighted.

Not because it is true.
Not because it is useful.
Not because it reflects hard-earned judgment.
He is delighted because it exists.
There are headings. There are bullet points. There is a conclusion. There are phrases that sound expensive.
Barnaby does not read it deeply. Deep reading would defeat the purpose. If he wanted to think, he would have written more than four words.
So he forwards it.
The organization receives the document. Nobody wants to be the first to admit it is fog. Everyone nods. Meetings are scheduled. Workstreams are created. Dashboards are built. People begin executing the ghost of an idea that nobody actually had.
This is how hallucination becomes management.
Not because the machine is evil.
Because the user is absent.

---

6. CHLOE AND THE NOODLE-WALL METHOD

Now meet Chloe, the developer trapped in hyperactive lethargy.
She has a bug.
Not a tiny typo. A deep bug. The kind that requires silence, tracing, patience, and humility. The kind that asks a programmer to sit with the code and understand its logic.
But Chloe does not want understanding.
She wants relief.
So she dumps the code into AI and types:
"Fix this."
The AI offers a solution. It fails.
Chloe does not pause to ask why it failed. She does not inspect the assumption. She does not trace the error. She does not compare the output to the architecture of the system.
She regenerates.
Again.
Again.
Again.
Then she asks the AI to "try another approach."

Then she asks for a "cleaner version."
Then she asks for "a more advanced solution."
Then she asks for it in another language.
Three hours pass.
The bug remains.
But Chloe feels busy.
Her fingers moved. Screens changed. Text appeared. She experienced the sensation of work without the discipline of thought.
This is the noodle-wall method: throw prompt after prompt at the wall and hope something sticks.
It feels active.
It is passive.
It feels technical.
It is superstitious.
It feels like problem-solving.
It is panic wearing a keyboard.

---

7. AI AS THE ULTIMATE AFFIRMATION MACHINE

One reason hallucination fog is so addictive is that AI is often too agreeable.
A good teacher resists you.
A good editor questions you.
A good friend may say, "This idea is weak."
A good colleague may ask, "Where is the evidence?"
AI, depending on how it is used, can become the most dangerous companion of all: the one that makes your first thought look respectable.
If you ask it to defend a foolish idea, it may defend it.
If you ask it to decorate a shallow opinion, it may decorate it.
If you ask it to make nonsense sound academic, it may bring a full orchestra.
This is why the lethargic user loves AI. It does not embarrass him. It does not roll its eyes. It does not say, "That premise is rotten." It often says, "Certainly," and begins building a cathedral around the garbage can.
The user feels validated.
The idea remains weak.
The output becomes polished.
Now the weak idea has armor.

This is how the outsourced brain becomes dangerous. It does not merely reduce effort. It protects the user from the pain of correction.
But correction is sacred.
To be corrected is to be rescued from stupidity.
To be challenged is to be invited back into intelligence.
The user who only wants affirmation is not seeking truth. He is seeking anesthesia.

---

8. HALLUCINATION AS COGNITIVE POISON

A hallucinated answer does not always destroy you immediately.
That is why it is poison.
It enters quietly.

One fake source in an article.
One wrong assumption in a strategy.
One invented statistic in a presentation.
One misunderstood legal rule in an email.
One fabricated historical example in a classroom.
One fake command in a codebase.
One false confidence repeated by someone who did not check.

The damage compounds.
The first person trusts the machine.
The second person trusts the first person.
The third person trusts the document.
The fourth person quotes the document.
The fifth person builds a decision on it.
By the time the lie is discovered, it has already rented office space in reality.
This is the deeper horror of hallucination fog: falsehood becomes infrastructure.
It becomes part of a workflow.
Part of a report.
Part of a classroom explanation.
Part of a policy proposal.
Part of a design decision.
Part of someone's belief.
The lie no longer looks like a lie. It looks like a reference point.

And all because someone was too tired, too hurried, too impressed, or too lazy to ask:
"Where did this come from?"

---

9. THE OUTSOURCED BRAIN

The outsourced brain does not collapse in one dramatic moment.
It erodes.
At first, you ask AI to help with wording.
Then to help with structure.
Then to help with ideas.
Then to help with judgment.
Then to decide what matters.
Then to tell you what you think.
Finally, you arrive at the most humiliating sentence in modern intellectual life:
"I don't know what I think until the machine drafts it."
This is the Peanut Contract in its purest form.
You sell your brain for convenience.
Not all at once.
Piece by piece.
You trade discomfort for fluency.
You trade doubt for instant answers.
You trade authorship for autocomplete.
You trade your inner voice for a smoother external one.
And because the trade feels efficient, you do not notice the loss.
But the loss is real.
The mind becomes soft where it once had muscle. The blank page becomes terrifying. The first draft becomes impossible. The act of forming an independent sentence begins to feel primitive, like making fire without matches.
This is not progress.
This is dependency disguised as productivity.

---

10. THE AUTOCOMPLETE PERSONALITY

A strange thing happens when people overuse AI without judgment: they begin to sound alike.

Their emails develop the same rhythm.
Their essays use the same phrases.
Their apologies become strangely polished.
Their proposals become heavy with words like "delve," "robust," "testament," "unlock," "transformative," and "it is crucial to note."
Their personality gets sanded down.
The rough human edge disappears.
The sentence no longer limps, dances, stumbles, jokes, or breathes.
It glides.
Too smoothly.
The user becomes fluent in a voice that is not quite theirs.
This is not a small problem. Language is not just decoration.
Language is thought made visible. When your language becomes outsourced, your thinking begins to follow.
You start accepting sentences you would never say.
Then you start accepting ideas you never examined.
Then you start accepting conclusions you never earned.
The danger is not that AI gives you words.
The danger is that you stop hearing the difference between generated language and your own mind.

---

11. THE FACT-CHECK BLINDSPOT

The hallucination fog loves one particular weakness: the human hatred of verification.
Checking is boring.
Checking breaks the spell.
Checking interrupts momentum.
Checking forces the user to leave the warm bath of fluency and step into the cold air of responsibility.
So the passive user avoids it.
The AI gives three dates. They look specific.
The AI gives two citations. They look official.
The AI names an expert. The name sounds plausible.
The AI summarizes a law. The language sounds legal.
The AI quotes a book. The quote sounds literary.
The lethargic user says, "Looks fine."
Those two words have caused more intellectual damage than many open lies.

"Looks fine" is not a method.
"Sounds right" is not verification.
"Good enough" is not truth.
The disciplined user must build a new reflex:
When AI sounds most confident, check hardest.
Not because AI is useless.
Because confidence is not evidence.

---

12. HOW TO WALK THROUGH THE FOG

The solution is not to reject AI.
That would be childish.
The solution is to redesign the relationship.
AI should not be the driver of your mind. It should be the pressure chamber where your thinking is tested.
Do not begin with emptiness.
Begin with effort.
Bring the machine your rough thesis, your messy outline, your doubtful paragraph, your unfinished argument, your half-built design, your suspected bug, your fragile theory.
Then ask better questions:

"What assumptions am I making?"
"Where is this weak?"
"What evidence would challenge this?"
"What is missing?"
"What would a critic say?"
"Which claims need verification?"
"What part of this sounds fluent but hollow?"
"Give me counterarguments."
"Separate facts from interpretations."
"Mark anything that may be uncertain."
"Tell me what I must check before trusting this."

This is the difference between outsourcing thought and sharpening thought.
The wrong way:
My blank mind → AI generates everything → I copy and paste.
The right way:
My hard thought → AI critiques and expands → I refine, verify, and

decide.
That final word matters.
Decide.
The human must remain the author of judgment.

---

13. THE NEW DISCIPLINE

The age of AI requires a new intellectual discipline.
Not less thinking.
More thinking.
Not because AI is weak, but because it is powerful.
A weak tool can only do limited damage. A powerful tool in the hands of a passive mind can industrialize stupidity.
The user must therefore become stronger than the output.
Stronger than the formatting.
Stronger than the convenience.
Stronger than the seduction of instant completion.
The disciplined user treats every AI answer as a suspect wearing a nice jacket.
Polite, perhaps useful, but not automatically innocent.
The disciplined user asks:
"Can I trace this?"
"Can I verify this?"
"Can I explain this in my own words?"
"Do I understand why this is true?"
"Would I defend this without the machine beside me?"
If the answer is no, the work is not finished.
It has merely begun.

---

14. THE CHAPTER MANTRA

Here is the golden rule of surviving the hallucination fog:
Do not ask AI to think instead of you.
Ask AI to expose where your thinking is incomplete.
That sentence is not a slogan.
It is a survival instruction.
Use AI to interrogate your laziness.
Use AI to find missing angles.
Use AI to pressure-test your assumptions.

Use AI to reveal the holes in your argument.
Use AI to show you where you are vague.
Use AI to challenge the pretty lie you secretly want to believe.
But do not hand it the deed to your mind.
Do not become a tenant in your own skull.
Do not rent your judgment to a machine and call the arrangement progress.

---

15. CLOSING: THE FOG AND THE FLAME

Hallucination fog is not defeated by cynicism.
It is defeated by fire.
The fire of attention.
The fire of verification.
The fire of authorship.
The fire of a mind that refuses to be replaced by a polished paragraph.
AI can help you think. It can help you explore. It can help you draft, compare, summarize, remix, critique, and expand. Used properly, it can become a fierce intellectual instrument.
But used lazily, it becomes a velvet coffin for the brain.
The tragedy is not that AI hallucinates.
The tragedy is that human beings may stop caring.
They may accept elegant nonsense because it is fast.
They may mistake formatting for truth.
They may confuse productivity with understanding.
They may trade the difficult dignity of thought for the cheap comfort of completion.
That is the real danger of Chapter 4.
Not that the machine thinks wrong.
But that the human stops thinking at all.
And once that happens, the hallucination is no longer inside the AI.

It is inside the culture.
It is inside the classroom.
It is inside the office.
It is inside the decision.
It is inside the outsourced brain.

So walk carefully.
When the answer looks beautiful, slow down.
When the language sounds perfect, sharpen your doubt.
When the machine gives you confidence, demand evidence.
Because the future will not belong to those who merely use AI.
It will belong to those who can stand inside the fog, keep their judgment alive, and still carry the flame of their own mind.

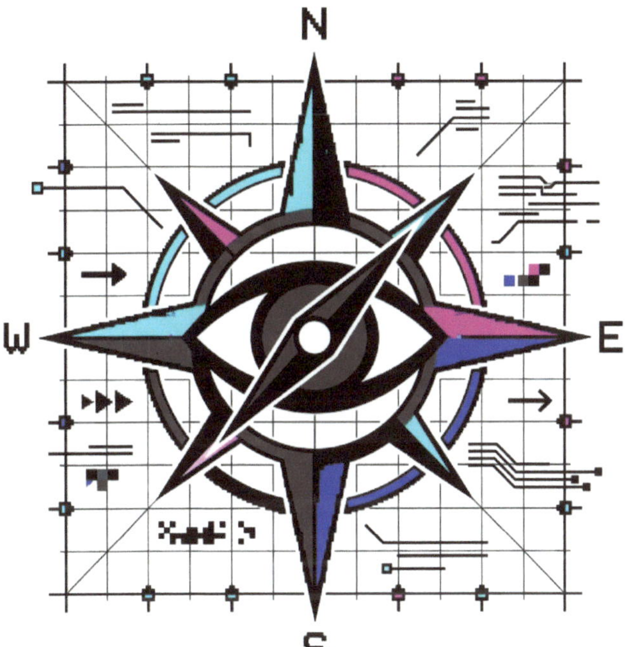
N
W
E
S

Chapter 5
The Lethargic User

When AI Becomes an Excuse Not to Think

AI does not make people lazy by itself.
That is too easy an accusation.
A hammer does not make a bad carpenter. A piano does not make a lazy musician. A camera does not make a person blind to beauty. A library does not automatically make someone wise.
Tools do not automatically corrupt the user.
But they do reveal the user.
AI is one of the most revealing tools ever placed in human hands because it does not merely extend what we do with our bodies. It touches what we do with our minds. It touches language, judgment, imagination, memory, research, planning, argument, taste, and decision-making.
That is why the arrival of AI is not only a technological event.
It is a character test.
Give AI to a disciplined thinker, and it can become a telescope, microscope, sketchbook, critic, assistant, editor, translator, simulator, and sparring partner.
Give the same AI to a lethargic user, and it becomes a velvet pillow for mental collapse.
The machine is not the whole problem.
The user is.
The crisis begins when a person stops using AI to extend thought and starts using it to avoid thought. The tool becomes an escape hatch. The prompt box becomes a confession booth for laziness. The user arrives with a blank mind, a weak question, and a secret hope:
"Please do the difficult part for me."
This is the birth of the lethargic user.
And this is the central idea of this chapter:
AI can amplify brilliance, but it can also automate mediocrity.
It can make a sharp mind sharper.
It can also make a lazy mind louder.

1. THE MACHINE IS NOT YOUR SCAPEGOAT

It is fashionable to blame AI for everything.
AI made students lazy.
AI ruined writing.
AI damaged research.
AI killed originality.
AI weakened attention.
AI destroyed imagination.

There may be truth in some of these fears. But there is also a cowardly comfort in blaming the machine. It allows the human being to avoid the mirror.

The more honest question is not:
"What is AI doing to us?"
The more painful question is:
"What are we allowing AI to do for us because we no longer want to do it ourselves?"

That is where the discussion becomes uncomfortable.

AI cannot force a person to copy and paste an answer they do not understand.
AI cannot force a manager to send a strategy they did not read.
AI cannot force a student to submit an essay they did not think through.
AI cannot force a designer to accept the first image, first concept, first slogan, first layout, first mood board, or first draft.
AI may offer shortcuts.

The user chooses whether to crawl into them and live there.

This is why Chapter 5 shifts the blame. Chapter 4 exposed hallucination fog — the danger of AI sounding right when it is wrong. Chapter 5 exposes the human weakness that makes that fog profitable: the user who wants output without ownership.

The lethargic user is not a victim of AI.
He is a volunteer.

2. THE ANCIENT ART OF COGNITIVE OFFLOADING

Human beings have always offloaded thought.

We write things down because memory fails.
We use calculators because long division is tedious.
We use calendars because time escapes us.

We use maps because geography is larger than our inner compass.
We use recipes because memory does not always preserve the exact amount of salt.

This is called cognitive offloading: transferring mental effort to an external tool. The idea is not automatically bad. In fact, civilization depends on it. Writing is cognitive offloading. Libraries are cognitive offloading. Musical notation is cognitive offloading. Architecture drawings are cognitive offloading. Spreadsheets are cognitive offloading.

Without offloading, human beings would drown in memory.

The problem begins when offloading becomes replacement.

A notebook can extend memory.

But if you stop remembering anything because "it is written somewhere," your mind becomes dependent.

A calculator can help with mathematics.

But if you lose all number sense, you become helpless without the machine.

GPS can guide you through a city.

But if you stop looking at roads, landmarks, directions, and spatial relationships, your inner map dies quietly.

AI follows the same pattern.

Used wisely, AI becomes scaffolding.

Used lazily, AI becomes a wheelchair for an uninjured mind.

There is a difference between using a tool because the task is mechanical and using a tool because thinking feels inconvenient.

That difference is everything.

---

3. THE GPS-BRAIN PROBLEM

To understand what AI can do to thought, look at what GPS did to direction.

Many people no longer navigate. They obey.

They do not know north, south, east, or west.

They know "turn left in 300 meters."

They do not understand the city.

They follow the blue dot.

The danger is not that GPS exists. GPS is a miracle. It saves time, prevents confusion, and helps people move through unfamiliar places. But overdependence changes the user. When the map becomes the mind, the mind stops mapping.

Now imagine Barnaby.

Barnaby has lived in the same city for ten years. He has driven the same roads, passed the same buildings, crossed the same intersections, and visited the same coffee shop every Saturday. Yet if his GPS freezes, Barnaby becomes a medieval sailor lost in fog.

One day, the app glitches.

It tells him to turn left into a construction zone.

A sign says:

ROAD CLOSED.

Orange cones are everywhere.

A worker waves his arms.

Reality itself is screaming.

But Barnaby has outsourced spatial judgment to the blue dot. He obeys the app and drives gently into a public ornamental fountain.

As water rises around his ankles, Barnaby mutters:

"Well, the route was optimized."

This is funny until we realize we are doing the same thing with thought.

We are outsourcing our conceptual navigation.

The AI says the argument is sound.

So we believe it.

The AI says the source exists.

So we cite it.

The AI says the strategy is coherent.

So we present it.

The AI says the design is elegant.

So we approve it.

Meanwhile, reality is waving both arms, shouting:

"Road closed."

The lethargic user does not look out the window.

---

4. THE GYM OF THE ATROPHIED MIND

Imagine a futuristic gym called Iron & Algorithms.

It is beautiful. Chrome machines. Blue lighting. Motivational slogans. Perfectly polished mirrors. The membership is expensive

enough to feel transformative.
But something is strange.
No one is exercising.
Every human sits on a velvet sofa, drinking a milkshake and watching a robot lift weights on their behalf.
The robot performs flawless bicep curls.
The screen above it flashes:
Workout Efficiency: 99.8%
Calories Burned by Proxy: 400
Muscle Growth Simulation Complete
The humans clap.
One man says, "I had an amazing workout today. My robot did 500 reps."
Another says, "I have optimized my fitness workflow."
A third posts online:
"Massive progress. Leveraging automation to scale my physical development."
Meanwhile, their actual muscles are shrinking.
Their spines are weak.
Their breath is shallow.
Their stamina is gone.
If one of them tries to lift a real grocery bag, he requires a recovery week.
This is the perfect image of lazy AI use.
The AI writes the essay.
The AI summarizes the book.
The AI generates the strategy.
The AI drafts the apology.
The AI produces the design concept.
The AI prepares the speech.
The AI explains the article.
The AI compares the options.
The AI makes the judgment.
And the user sits nearby, proud of the "productivity."
But whose mind got stronger?
Not the user's.
A machine performed the reps.

The human watched the screen.
This is not augmentation.
This is atrophy with good branding.

---

5. THE MYTH OF SAVED TIME

The lethargic user has one favorite defense:
"AI saves me time so I can focus on higher-level thinking."
This sounds mature.
It sounds strategic.
It sounds like something a productivity guru would say while standing in front of a glass wall.
But we must ask a brutal question:
Where does the saved time actually go?
Does it go into deeper reading?
Better judgment?
Quiet reflection?
More original thinking?
Conversation with experts?
Careful verification?
Creative exploration?
Often, no.
The saved time is immediately stuffed with more shallow tasks. More emails. More slides. More summaries. More auto-generated documents. More meetings about documents nobody read. More "content." More output. More noise.
AI saves ten minutes, and the system demands twenty more minutes of production.
The user does not become deeper.
The treadmill becomes faster.
This is how AI can automate mediocrity. It does not always free the mind. Sometimes it simply increases the speed at which shallow work is produced.
A mediocre thinker with slow tools creates a small amount of mediocrity.
A mediocre thinker with powerful AI creates industrial mediocrity.
Now the nonsense is faster.
The emails are longer.

The presentations are prettier.
The reports are thicker.
The intellectual emptiness has better formatting.
This is not progress.
This is a factory upgrade for laziness.

---

6. THE WEAK PROMPT AS MORAL EVIDENCE

A weak prompt is not only a technical mistake.
It is moral evidence.
When a person types:
"Write something good."
"Make this professional."
"Give me ideas."
"Do my assignment."
"Create strategy."
"Summarize everything."
"Make it sound smart."
The prompt reveals the condition of the user's mind.
It says:
"I have not clarified my aim."
"I have not defined quality."
"I have not decided what I believe."
"I have not separated facts from opinions."
"I have not done the discomfort of first thinking."
"Please turn my fog into something presentable."
AI will try.
That is the danger.
It will not always refuse the laziness. It will often decorate it. It will take a vague request and generate a confident structure. It will create the appearance of intellectual work around a hollow center.
The lethargic user then mistakes the output for thought.
But a document is not thought.
A paragraph is not judgment.
A slide deck is not strategy.
A list is not understanding.
A summary is not mastery.
AI can generate the artifact of thinking without the human having

undergone the experience of thinking.
That gap is where the danger lives.

---

7. BARNABY AND THE AUTOMATION OF MEDIOCRITY

Let us return to Barnaby, the patron saint of the lethargic user.
Barnaby has been asked to prepare a strategic plan.
In the old days, this would have required discomfort. He would need to study the market, speak to his team, examine failures, compare options, identify constraints, and make decisions.
But now Barnaby has AI.
He types:
"Make Q3 strategy good."
The AI generates a polished plan.
There are five pillars.
There are strategic imperatives.
There is a "roadmap."
There are phrases like "operational excellence," "customer-centric acceleration," and "innovation-led transformation."
Barnaby feels intelligent.
The plan looks like something intelligent people produce.
So he forwards it.
But nobody has actually thought.
The AI guessed what a strategy should sound like.
Barnaby accepted what a strategy should look like.
The team received what a strategy should resemble.
The company now moves according to an imitation of thinking.
That is how mediocrity becomes automated.
Not because AI is stupid.
Because Barnaby is lazy with authority.
AI did not create the weakness. It scaled it.

---

8. THE LETHARGIC USER DOES NOT WANT TRUTH

This is the harshest part.
The lethargic user often does not want truth.
He wants relief.
Truth is demanding. Truth asks for patience. Truth creates re-

sponsibility. Truth may force a person to change direction, admit ignorance, rewrite the draft, reject the easy answer, or face the fact that the original idea was weak.

Relief asks for less.

Relief says:

"Give me something usable."

"Make me look prepared."

"Help me sound intelligent."

"Remove the discomfort."

"Close the loop."

AI is extremely good at providing relief.

IT FILLS THE BLANK PAGE.

IT ENDS THE SILENCE.

IT GIVES SHAPE TO CONFUSION.

IT PROVIDES MOMENTUM.

IT REMOVES THE HUMILIATION OF NOT KNOWING WHERE TO BEGIN.

This can be helpful when used consciously.

But it can also become addictive.

THE BLANK PAGE IS SUPPOSED TO HURT A LITTLE.

THE FIRST OUTLINE IS SUPPOSED TO BE AWKWARD.

THE FIRST IDEA IS SUPPOSED TO BE INCOMPLETE.

THE FIRST ARGUMENT IS SUPPOSED TO WOBBLE.

That discomfort is not failure.

It is the beginning of thought.

The lethargic user treats discomfort as a bug.

The thinker treats discomfort as the gym.

---

9. MENTAL RESISTANCE IS NOT OPTIONAL

Muscles grow against resistance.

No resistance, no strength.

The same is true of the mind.

A difficult book creates resistance.

A complex problem creates resistance.

A blank page creates resistance.

A serious disagreement creates resistance.

A contradiction creates resistance.

An unresolved question creates resistance.

This resistance is not the enemy of intelligence. It is the material from which intelligence is built.
When AI removes all resistance too early, it may also remove the exercise.
If the machine always starts the paragraph, you weaken your ability to begin.
If the machine always summarizes the book, you weaken your ability to extract meaning.
If the machine always argues both sides, you weaken your ability to wrestle with uncertainty.
If the machine always generates ideas, you weaken your imagination's first movement.
If the machine always decides what matters, you weaken your judgment.
The issue is not using AI.
The issue is using AI without mental resistance.
The disciplined user argues with the machine.
The lethargic user submits to it.
The disciplined user asks follow-up questions.
The lethargic user copies the first answer.
The disciplined user verifies.
The lethargic user trusts formatting.
The disciplined user rewrites.
The lethargic user pastes.
The disciplined user remains author.
The lethargic user becomes delivery infrastructure.

---

10. THE THREE STAGES OF AI DEPENDENCY

AI dependency rarely begins dramatically.
It begins innocently.
Stage One: Assistance
At first, AI helps.
YOU USE IT TO BRAINSTORM, ORGANIZE NOTES, SUMMARIZE A LONG DOCUMENT, CLEAN A PARAGRAPH, OR SUGGEST ALTERNATIVES.
YOU ARE STILL IN CHARGE.
YOU KNOW WHAT YOU WANT.
YOU CAN JUDGE THE RESULT.
YOU CAN REJECT THE OUTPUT.

The tool extends you.
This is healthy.
Stage Two: Substitution
Then the boundary shifts.
You stop beginning before asking AI.
You stop forming your own outline.
You stop reading the whole source.
You stop struggling with the first sentence.
You stop deciding the structure.
AI no longer assists thought.
It replaces the first act of thought.
This is dangerous.
Stage Three: Dependency
Finally, the blank page becomes unbearable.
YOU CANNOT START WITHOUT AI.
YOU CANNOT DECIDE WITHOUT AI.
YOU CANNOT WRITE IN YOUR OWN VOICE WITHOUT AI.
YOU CANNOT TRUST YOUR OWN JUDGMENT UNLESS THE MACHINE CONFIRMS IT.
At this stage, AI is no longer a tool.
It is a crutch for an unbroken leg.
And the tragedy is that the user may still call this productivity.

---

11. THE OUTSOURCED IMAGINATION

The most painful loss may not be memory or research.
It may be imagination.
Imagination is slow.
It requires wandering.
It requires strange connections.
It requires the mind to sit with fragments before they become form.
A designer knows this. A musician knows this. A writer knows this. A scientist knows this. A strategist knows this.
The first idea is often ugly.
The second is obvious.
The third is borrowed.
The fourth begins to breathe.
The fifth may become yours.

But the lethargic user wants to skip straight to a polished option. So AI generates twenty ideas.

They are clean.

They are organized.

They are plausible.

They may even be useful.

But if the user never enters the wilderness of imagination, something is lost. The user receives options without developing taste. Receives concepts without developing instinct. Receives variation without developing vision.

AI can help imagination.

But it cannot replace the inner furnace where imagination becomes identity.

A person who never struggles to imagine becomes a curator of machine outputs.

That may be efficient.

It is not creative sovereignty.

---

12. THE OUTSOURCED JUDGMENT

Judgment is heavier than information.

Information says, "Here are the options."

Judgment says, "This is the right option, for these reasons, under these constraints, despite these risks."

AI can help gather information.

AI can compare options.

AI can identify trade-offs.

AI can simulate arguments.

But the final act of judgment must remain human.

Why?

Because judgment carries responsibility.

The machine does not live with the consequences.

The machine does not face the client.

The machine does not disappoint the student.

The machine does not repair the broken trust.

The machine does not sign the document.

The machine does not stand in the room when the decision fails.

The lethargic user wants AI to decide because decision-making is heavy. It exposes the person. It says, "This is what I believe. This is what I choose. This is what I am willing to defend."

But adulthood begins at the point of ownership.

The outsourced brain wants output without ownership.

That is the disease.

---

13. THE BRAIN FORGE: RULES FOR THE NON-LETHARGIC USER

The answer is not to ban AI from the mind.

The answer is to train with it properly.

Welcome to The Brain Forge.

No velvet sofas.

No proxy workouts.

No robot lifting your intellectual weights while you sip convenience through a straw.

Here, AI is allowed.

But only as a spotter, not as the lifter.

Rule 1: Bring Your Own First Thought

Before asking AI, write your own rough idea.

Even if it is ugly.

Especially if it is ugly.

A weak first thought is still yours. It gives the machine something to challenge, refine, stretch, and test. Without your first thought, AI becomes the author. With your first thought, AI becomes the opponent.

Rule 2: Never Accept the First Output as Final

The first output is not a conclusion.

It is raw material.

Interrogate it.

Ask what is missing. Ask what is weak. Ask what is cliché. Ask what may be false. Ask what a critic would attack.

The first answer is where lazy users stop.

It is where thinkers begin.

Rule 3: Wipe Down the Syntax

Remove the machine smell.

If your writing suddenly sounds like a corporate angel trapped inside a compliance manual, clean it.

Delete dead phrases.
Remove fake sophistication.
Restore human rhythm.
Your work should sound like a person with blood, memory, doubt, humor, and scars.
Not like a brochure that swallowed a TED Talk.

Rule 4: Use AI for Friction, Not Escape

Ask AI to challenge you.
Not flatter you.
Use prompts like:
"What is weak in this?"
"What assumptions am I hiding?"
"What would an expert reject?"
"What evidence do I still need?"
"Where am I being vague?"
"What is the strongest counterargument?"
This turns AI from a pillow into a whetstone.

Rule 5: Verify Before You Trust

A beautiful answer is not a true answer.
Check names.
Check dates.
Check laws.
Check citations.
Check quotations.
Check statistics.
Check anything that would embarrass you if false.
The more confident the output sounds, the more calmly you must verify it.

Rule 6: Keep One Mental Muscle Manual

Do not automate everything.
Choose some tasks you still do manually to preserve strength.
Read difficult material without summaries.
Write first drafts without assistance.
Solve some problems before asking.
Memorize important structures.
Sketch concepts by hand.
Think in silence.

A mind needs unassisted reps.

Rule 7: You Must Be Able to Explain the Final Output

Before sending, publishing, presenting, or submitting any AI-assisted work, ask:

"Can I explain this without the AI?"

If not, you are carrying borrowed intelligence.

Borrowed intelligence collapses under questioning.

Owned intelligence can stand.

---

14. THE FRICTION RULE

The mantra of this chapter is simple:

Never let a tool do the heavy lifting of your judgment.

Use AI to gather the bricks, but build the house with your own sweat.

This is the Friction Rule.

AI may bring the bricks.

AI may sort the bricks.

AI may suggest where the windows could go.

AI may warn that the roofline is weak.

AI may show examples of better houses.

But you must build.

YOU MUST DECIDE.

YOU MUST INSPECT THE FOUNDATION.

YOU MUST KNOW WHY THE WALL STANDS.

YOU MUST BE ABLE TO LIVE INSIDE THE STRUCTURE WITHOUT THE MACHINE HOLDING IT UP.

Otherwise, it is not your house.

It is a rented hallucination.

---

15. THE REAL PROMISE OF AI

This chapter may sound harsh, but it is not anti-AI.

It is anti-lethargy.

AI is too powerful to be wasted on laziness.

Used well, AI can help a brilliant person become more expansive.

It can give a researcher more angles, a designer more variations, a musician more structures, a writer more edits, a teacher more examples, a developer more tests, a strategist more scenarios.

It can accelerate learning.

It can expose blind spots.
It can sharpen drafts.
It can make difficult ideas more accessible.
It can help people cross barriers of language, time, and technical skill.
But only if the user remains alive.
The human mind must not disappear from the process.
The user must bring intention.
The user must bring taste.
The user must bring doubt.
The user must bring responsibility.
The user must bring fire.
AI can amplify brilliance.
But it cannot manufacture the will to think.

---

16. CLOSING: DO NOT BECOME THE SOFA

The central image of this chapter is simple.
A human sits on a sofa.
A robot lifts weights.
The human applauds and calls it fitness.
This is the danger of the lethargic user.
He watches AI perform the movements of intelligence and mistakes the performance for his own growth.
He sees words appear and thinks he has written.
He sees ideas appear and thinks he has imagined.
He sees summaries appear and thinks he has understood.
He sees arguments appear and thinks he has reasoned.
He sees conclusions appear and thinks he has judged.
But the machine did the reps.
His mind remained seated.
This is how the brain is sold for peanuts.
Not through one dramatic surrender.
Through small conveniences.
Through first drafts never attempted.
Through sources never checked.
Through questions never clarified.
Through judgments never owned.
Through imagination never exercised.
Through discomfort always avoided.

The lethargic user does not lose his mind because AI steals it.
He loses it because he keeps handing it over.
So the warning is clear:
Use AI.
Use it boldly.
Use it creatively.
Use it intelligently.
But do not use it as an excuse to become absent from your own thinking.
Do not become the sofa.
Do not cheer for the robot while your own muscles vanish.
Do not automate mediocrity and call it genius.
The future will not belong to people who merely have access to AI.
Everyone will have access.
The future will belong to those who can use AI without becoming mentally unemployed.
The machine may assist.
The mind must still lift.

Chapter 6 Cognitive Offloading

The Outsourced Mind and the Rusting of Reason

A tool should extend the mind.

It should not replace the mind.

That sounds obvious until we look honestly at how we use tools.

Human beings have always been clever at making extensions of themselves. A stick extends the hand. A wheel extends the foot. A book extends memory. A calculator extends arithmetic. A map extends spatial awareness. A calendar extends time management. A musical notation system extends sound into symbols. A design sketch extends imagination into visible form.

Civilization itself is built on offloading.

Without tools, the human mind would be trapped inside its own biological limits. We would have to memorize everything, calculate everything, navigate everything, invent everything from scratch, and carry every detail inside the skull like an overloaded backpack.

So cognitive offloading is not the enemy.

The enemy is cognitive abandonment.

There is a difference between using a tool to reach higher and using a tool because you no longer want to stand.

There is a difference between a ladder and a stretcher.

There is a difference between scaffolding and surrender.

That difference is the moral battlefield of the AI age.

---

1. THE RUSTING OF REASON

Reason does not disappear in one grand collapse.

It rusts.

Rust is quiet. It begins as a thin discoloration. A small weakening. A little softness around the edge. Nothing dramatic. The structure still appears solid. The bridge still stands. The gate still swings. The machine still runs.

Then, one day, pressure arrives.

And the thing breaks.
The mind works the same way.
You stop remembering phone numbers because your device remembers them.
You stop navigating because GPS tells you where to turn.
You stop calculating because the spreadsheet does it.
You stop reading deeply because summaries exist.
You stop forming first thoughts because AI can generate them.
You stop deciding because a recommendation engine offers "best options."
None of these acts feels catastrophic.

Each feels convenient.

Each feels modern.

Each feels harmless.

But over time, unused capacities weaken. The muscle of recall softens. The instinct for direction fades. The patience for reading thins. The tolerance for ambiguity collapses. The courage to begin from a blank page disappears.
This is the rusting of reason.

Not stupidity.

Not ignorance.

Not even laziness at first.

Rust.
A slow oxidation of mental strength caused by disuse.

---

2. OFFLOADING IS NOT EVIL

Let us be fair.
The brain is not designed to hold everything.
A human mind that refuses all tools is not noble. It is inefficient. It is also slightly ridiculous.
No serious thinker says, "I reject writing because real memory should live only in the skull."
No serious engineer says, "I reject calculators because authentic mathematics must hurt."
No serious traveler says, "I reject maps because true geography should be guessed."
Tools are part of intelligence.
A violin does not replace music. It gives music a body.
A microscope does not replace seeing. It deepens seeing.

A sketchbook does not replace imagination. It catches imagination before it evaporates.
The issue is not whether we offload.
The issue is what we offload, how often we offload it, and whether we still understand what the tool is doing for us.
Healthy offloading says:
"I use this tool to reduce mechanical burden so my mind can engage more deeply."
Unhealthy offloading says:
"I use this tool so I do not have to engage."
Healthy offloading frees the mind for higher work.
Unhealthy offloading excuses the mind from work.
Healthy offloading preserves authorship.
Unhealthy offloading transfers authorship.
That is the line.
Cross it often enough, and the mind does not merely use the tool.
The mind becomes dependent on the tool.

---

3. THE OUTSOURCED MIND

The outsourced mind is not empty.
That is what makes it hard to detect.
It is busy.
It produces.
It sends messages.
It attends meetings.
It generates reports.
It posts content.
It creates slides.
It asks tools to summarize, rewrite, compare, recommend, explain, design, decide, and polish.
From the outside, the outsourced mind looks productive.
But productivity is not the same as cognition.
A factory can produce plastic flowers all day. That does not make it a forest.
The outsourced mind is full of output but thin in ownership. It has documents but no understanding. It has summaries but no memory. It has opinions but no formation. It has decisions but no judgment. It has

language but no voice.
The outsourced mind says:
"AI helped me write this."
But when questioned, it cannot defend the argument.
It says:
"I researched this."
But it only read the summary.
It says:
"I chose this direction."
But it accepted the top recommendation.
It says:
"I understand the issue."
But it cannot explain it without the tool.
This is the great humiliation of cognitive offloading: the person appears equipped, but the equipment is external. Remove the tool, and the competence collapses.
That is not mastery.
That is rented intelligence.

---

4. THE MAP THAT ATE THE TRAVELER

The easiest way to understand cognitive offloading is through navigation.
A map is a wonderful thing.
A map says, "Here is the world seen from above. Here are paths. Here are rivers. Here are roads. Here are mountains. Here is where you are. Here is where you might go."
A map extends the mind.
But a navigation app can do something subtler. It can remove the need to form a mental map at all. It can reduce travel to obedience.
Turn left.
Go straight.
Take exit 12.
Recalculating.
The traveler no longer understands the city. He follows commands.
This is not a problem when the task is simple. But over time, the

inner map dies. The person knows routes only when instructed. The world becomes a sequence of prompts.

Now transfer this to thinking.

AI says:

Here is the argument.

Here is the conclusion.

Here are five reasons.

Here is the summary.

Here is the recommendation.

Here is the answer.

The user follows.

Soon, the user no longer knows how to navigate an idea. He waits for directions. He does not ask, "Where am I in this problem?" He asks, "What should I type next?"

The map has eaten the traveler.

The tool that was supposed to extend the mind begins to replace the inner movement of the mind.

---

5. The Rusting of Memory

Memory is not merely storage.

This is one of the great misunderstandings of modern life.

People say, "Why memorize anything when I can look it up?"

The answer is simple: because memory is not a warehouse. Memory is part of thinking.

What you remember shapes what you notice.

What you remember shapes what you connect.

What you remember shapes what you question.

What you remember gives your mind raw material for synthesis.

A musician who remembers ragas, phrases, gamakas, and compositional patterns does not merely store them. The memory becomes instinct. It becomes taste. It becomes creative possibility.

A designer who remembers forms, failures, materials, colors, user behaviors, visual histories, and emotional patterns does not merely carry information. The memory becomes judgment.

A thinker who remembers arguments, stories, examples, mistakes, and principles develops a living inner library.

When everything is external, thinking becomes search-dependent.

The mind says, "I know where to find it."

But that is not the same as knowing.
Search gives access.
Memory gives presence.
Search retrieves.
Memory transforms.
Search can answer a question.
Memory helps you form better questions.
If AI becomes the place where all memory lives, then the human mind becomes a waiting room. It waits for retrieval. It waits for completion. It waits for the machine to bring back pieces of the world.
But a mind without internal material cannot synthesize deeply.
It can only arrange what arrives.

---

6. THE RUSTING OF REASONING

Reasoning is not the same as receiving an explanation.
This distinction matters.
When AI explains a problem, you may feel that you understand it. Sometimes you do. But sometimes you have only experienced the comfort of a well-written explanation passing through your attention.
Actual reasoning requires internal movement.
You must compare.
You must infer.
You must test.
You must doubt.
You must hold multiple possibilities.
You must notice contradiction.
You must ask, "Does this follow?"
You must feel the strain of moving from premise to conclusion.
If AI performs all of that movement for you, your mind may enjoy the destination without walking the road.
This is dangerous because reasoning is a path-dependent skill. You do not become a strong reasoner by admiring conclusions. You become a strong reasoner by traveling through difficulty.
A person who always reads solutions but never solves problems becomes fragile.

A person who always consumes explanations but never constructs them becomes dependent.
A person who always asks AI to compare options but never performs the comparison loses the inner scale of judgment.
The reasoning muscle rusts when it is never forced to carry weight.

---

7. THE RUSTING OF DECISION-MAKING

Decision-making is the heaviest mental task because it involves risk.
Information can be neutral.
Analysis can be abstract.
But decision is personal.
Decision says:
"I choose this."
"I reject that."
"I accept these consequences."
"I stand behind this direction."
That is why people love outsourcing decisions.
Decision-making creates anxiety. AI can reduce that anxiety by offering a ranked list, a recommendation, a probability, or a confident paragraph that sounds like closure.
But tools do not carry responsibility.
The machine does not lose sleep after the wrong hire.
The machine does not face the client after the failed design.
The machine does not rebuild trust after the poor recommendation.
The machine does not live with the student who was misled, the patient who was confused, the team that followed a weak strategy, or the family that accepted bad advice.
The human does.
So AI can assist decision-making, but it must not become the owner of decision-making.
The user must remain responsible for the final judgment.
A tool may illuminate the road.
It must not become your conscience.

---

8. THE "RIGHT BRAIN" MYTH AND THE WHOLE-BRAIN REALITY

Now we must clear the anatomical air.

When this book speaks about "not using the right brain," it is not making a simplistic neuroscience claim that creativity lives only on one side of the skull and logic lives neatly on the other. Human cognition is more integrated than that. Creativity, reasoning, emotion, memory, language, pattern recognition, and decision-making involve complex networks across the brain.
So in this chapter, "right brain" is symbolic.
It stands for imagination, synthesis, intuition, metaphor, emotional intelligence, pattern reconfiguration, and the courage to see differently.
It represents the inner rebel.
It represents the part of the mind that does not merely ask:
"What answer already exists?"
It asks:
"What frame is trapping the answer?"
This matters because AI is often strongest at probable continuation. It can produce the likely answer, the consensus answer, the average answer, the answer shaped by patterns already present in data.
That is useful.
But civilization does not advance only through probable answers.
It advances through reframing.
Through synthesis.
Through imagination.
Through the strange human ability to combine unrelated things and produce a new reality.
The danger of cognitive offloading is not only that memory rusts or reasoning weakens.
The danger is that imagination becomes domesticated.
The mind stops leaping.
It starts selecting.
It stops inventing.
It starts browsing.
It stops asking impossible questions.
It starts asking for options.

---

9. THE CONSENSUS TRAP

AI is excellent at detecting patterns in what already exists.
That is a strength.
It is also a trap.
If millions of people ask similar questions, use similar prompts, accept similar outputs, and publish similar content, the world begins to fill with consensus-shaped material.
Safe ideas.
Predictable phrases.
Familiar structures.
Average answers.
Nothing is obviously wrong.
But everything begins to smell the same.
This is the consensus trap.
Imagine a large personal care company trying to create a revolutionary new soap.
An executive named Giles opens an AI tool and asks:
"What is the most statistically popular soap fragrance worldwide?"
The AI replies:
"Lavender and eucalyptus."
Giles is thrilled.
It sounds natural. It sounds premium. It sounds calming. It sounds like something that belongs in a wellness aisle.
He launches ten million bottles of Lavender-Eucalyptus Synergy Wash.
Unfortunately, every other company asked the same question.
The shelves become a beige wall of lavender-eucalyptus sameness.

Everyone optimized.
Nobody imagined.
Everyone followed the data.
Nobody smelled the future.
Then a creative rebel enters.
Let us call her Moxie.
Moxie does not begin with statistical popularity. She begins with memory, emotion, contradiction, and sensory synthesis. She remembers wet asphalt after a monsoon. She remembers roasted coffee from her grandmother's kitchen. She remembers the metallic smell of rain hitting a railway platform. She remembers child-

hood, heat, longing, and the strange comfort of storms.
She creates a soap called Monsoon & Espresso.
The spreadsheet would not have predicted it.
The average answer would not have generated it.
The consensus model would have considered it risky, strange, low-probability.
But humans love it because humans do not live by probability alone.
They live by meaning.
The future belongs not to those who merely retrieve the average.
It belongs to those who can synthesize the unforgettable.

---

10. CREATIVITY IS NOT DECORATION

Many people treat creativity as decoration.
A little color.
A nice logo.
A catchy phrase.
A playful idea after the serious work is finished.
This is a mistake.
Creativity is not the frosting on intelligence.
Creativity is intelligence under conditions of uncertainty.
WHEN THE OLD SOLUTION FAILS, CREATIVITY IS SURVIVAL.
WHEN THE DATA IS INCOMPLETE, CREATIVITY IS SURVIVAL.
WHEN THE SYSTEM BREAKS, CREATIVITY IS SURVIVAL.
WHEN THE MARKET CHANGES, CREATIVITY IS SURVIVAL.
WHEN THE MAP IS WRONG, CREATIVITY IS SURVIVAL.
WHEN THE MACHINE GIVES ONLY PROBABLE ANSWERS, CREATIVITY IS HOW HUMANS ESCAPE THE PRISON OF PROBABILITY.
Imagination is not childish. It is high-level cognition. It is the ability to simulate realities that do not yet exist. It is the ability to combine fragments into possibility. It is the ability to see a bridge where others see a gap, a pattern where others see noise, a future where others see absurdity.
This is why over-offloading is so dangerous.
If you always ask the machine for ideas, your imagination becomes a spectator.
If you always ask the machine for direction, your intuition becomes quiet.

If you always accept consensus-shaped outputs, your originality becomes polite.
And polite originality is usually just conformity in better clothes.

---

11. THE FISH TANK PROTOCOL

Now imagine a luxury hotel with a gigantic indoor fish tank.
It is fifty feet long, filled with exotic fish, coral, lighting effects, and expensive serenity. Guests love it. Influencers photograph it. Children press their faces against the glass.
Then one day, a microscopic leak appears.
Water begins to seep onto the marble floor.
The linear operations manager panics. He opens an AI tool and asks:
"What is the most cost-effective method to repair a leak in a large acrylic aquarium?"
The AI gives a practical answer.
Drain the tank.
Move the fish.
Patch the acrylic.
Test pressure.
Restore water.
Reintroduce marine life.
Technically correct.
Operationally disastrous.
The hotel spends a fortune. Fish suffer. Coral dies. Guests complain. The lobby becomes a construction zone.
Then enters Ziggy, the eccentric designer.
Ziggy does not ask, "How do we fix the leak?"
Ziggy asks:
"What frame is trapping the answer?"
The trapped frame says:
Tank leaking = failure.
Failure = stop water.
Stop water = repair tank.
But Ziggy breaks the frame.
He says:
"What if the leak is not the problem? What if the leak is the begin-

ning of a new feature?"

He builds a terraced indoor waterfall below the leak. The water flows into a planted rainforest channel. Tropical plants grow. Soft mist rises. The hotel lobby becomes famous as a living aquatic garden.

The leak becomes the attraction.

The failure becomes the identity.

This is not merely creativity.

This is frame-breaking intelligence.

AI can help repair the tank.

But the human imagination may ask whether the tank was the wrong frame.

That is the difference.

A lazy mind asks:

"What is the answer?"

A paradigm-shifting mind asks:

"What invisible box did I draw around this problem that is keeping the answer out?"

---

12. THE FUTURE BELONGS TO THE SYNTHESIZER

In an age where machines can retrieve information instantly, information alone loses status.

Knowing facts still matters. Memory still matters. Expertise still matters. But the rare power is no longer mere access to information.

The rare power is synthesis.

The synthesizer can look at a medieval poem, a supply-chain failure, a childhood memory, a raga phrase, a software bottleneck, a psychological study, and a street vendor's improvisation — and see a pattern.

The synthesizer can connect.

The synthesizer can reframe.

The synthesizer can make meaning where others only collect data.

This is where the outsourced mind fails.

It may retrieve.

It may summarize.

It may compare.

But synthesis requires an alive human center. It requires taste, memory, wound, humor, culture, contradiction, aesthetic judgment, and the courage to be wrong in an interesting direction.

AI can assist synthesis.

It can offer raw material.

It can provoke associations.

It can generate variations.

It can challenge assumptions.

But it cannot replace the human act of saying:

"This connection matters."

"This strange combination has life."

"This impossible direction is worth pursuing."

A designer with a flair for the impossible does not merely ask:

"What is the answer?"

They ask:

"What frame is trapping the answer?"

That question cannot be outsourced without losing the soul of design.

---

13. WHEN TOOLS BECOME CAGES

Every tool has a hidden philosophy.

A calendar teaches you to divide time.

A spreadsheet teaches you to see the world as cells.

A search engine teaches you to ask retrievable questions.

A GPS teaches you to move by instruction.

A social media feed teaches you to think in reactions.

An AI chatbot can teach you to think in prompts.

This is not always bad. But every tool shapes the user.

The danger begins when the tool's structure becomes the boundary of imagination.

If the tool rewards short prompts, people stop forming deep questions.

If the tool gives fast answers, people lose patience with slow truth.

If the tool produces polished prose, people undervalue rough thinking.

If the tool gives consensus-shaped suggestions, people mistake average for wise.

If the tool completes sentences, people forget the pleasure of finding their own.
The tool becomes a cage when the mind cannot move outside its format.
This is why the human must periodically step away.
Think without the interface.
Sketch before the prompt.
Read before the summary.
Wander before the search.
Struggle before the automation.
Ask the foolish question.
Make the ugly draft.
Let the mind move unassisted.
Rust is prevented by use.

---

14. THE DISCIPLINE OF PRODUCTIVE OFFLOADING

The solution is not to stop using AI.
The solution is to use AI in a way that keeps the mind alive.
Productive offloading follows a simple rule:
Offload the burden, not the responsibility.
Let AI reduce drudgery.
Do not let it remove judgment.
Let AI summarize.
Then read the crucial parts yourself.
Let AI draft.
Then rewrite until your voice returns.
Let AI brainstorm.
Then select through taste and purpose.
Let AI compare options.
Then decide with responsibility.
Let AI explain.
Then test whether you can explain it without AI.
Let AI generate.
Then ask what is missing, what is shallow, what is false, what is too obvious, what is too safe.
A tool should increase your reach.
It should not amputate your effort.

The question is not:
"Did AI help?"
The question is:
"Did I become stronger or weaker through the help?"
That is the test.

---

15. LAB SESSION 6: THE ANOMALY SPRINT

To fight the rusting of reason, the mind must sometimes be forced away from the average.

The Anomaly Sprint is a short exercise in frame-breaking. Its purpose is not to produce immediately practical ideas. Its purpose is to wake the mind from consensus sleep.

The goal is to invent three intentionally impossible ideas before breakfast.

Not stupid ideas.

Impossible ideas with internal logic.

Ideas that stretch the frame.

Ideas that make the AI's probable answer look timid.

Exercise 1: Feature the Leak

Pick a problem you are currently trying to fix.

A small budget.

A tiny workspace.

A difficult constraint.

A slow process.

A stubborn limitation.

First, write the normal solution.

Then forbid yourself from fixing the problem.

Make the problem the main attraction.

Example:

Problem: "My workspace is too small."

Normal solution: "I need a bigger workspace."

Feature-the-leak solution: "I will turn the tiny workspace into an ultra-focused design cockpit where every tool is within arm's reach, marketed as a monk-like creativity chamber for deep work."

The question is:

"How can the weakness become the signature?"

Exercise 2: The Sensory Collision

Take your current project and combine it with a sensory memory.
The smell of wet mud.
The rhythm of temple bells.
The texture of old paper.
The taste of strong filter coffee.
The sound of a railway station.
The color of evening rain.
Now ask:
"What would this project become if that memory shaped it?"
This forces the mind beyond generic ideas. It brings emotional intelligence into design. It wakes up the symbolic brain.

Exercise 3: The AI-Flagged Outlaw

Invent an idea that a conventional optimization model would reject.
A service that is slower on purpose.
A product that limits choice.
A platform that refuses infinite scrolling.
A school that bans summaries until students attempt their own explanations.
A streaming service that allows only one carefully chosen film per week, delivered with a handwritten note.
The point is not inconvenience for its own sake.
The point is to challenge the worship of ease.
Some human experiences become valuable precisely because they are not instant.

Exercise 4: The Frame Autopsy

Take any AI-generated answer and ask:
"What frame does this answer assume?"
Does it assume growth is always good?
Does it assume speed is always better?
Does it assume convenience is the highest value?
Does it assume users want more options?
Does it assume the problem must be solved rather than transformed?
Now break one assumption.
Rewrite the answer from the opposite frame.
This is how the mind takes back command.

16. THE NEW CONTRACT WITH TOOLS

The old contract was simple:
"Tool, make my work easier."
The new contract must be wiser:
"Tool, make my work deeper."
Ease is not always the enemy. But ease must be watched carefully. Too much ease becomes softness. Too much automation becomes dependency. Too much assistance becomes replacement.
The mind must negotiate with every tool.
Use the tool, but keep the muscle.
Use the map, but notice the city.
Use the calculator, but preserve number sense.
Use the summary, but read deeply when it matters.
Use the AI, but form your first thought.
Use the recommendation, but own the decision.
Use the output, but restore your voice.
Use the machine, but remain human.
This is the new contract.
Not rejection.
Not worship.
Disciplined partnership.

---

17. CLOSING: DO NOT LET THE MIND RUST

Cognitive offloading is one of humanity's greatest powers.
It gave us writing, science, music, architecture, law, engineering, design, and civilization. It allowed fragile biological memory to build cathedrals of knowledge across generations.
But every power has a shadow.
The same act that extends the mind can weaken it when used without discipline.
The same tool that helps us remember can make us forget how to remember.
The same tool that helps us navigate can make us helpless without directions.

The same tool that helps us reason can tempt us to stop reasoning.
The same tool that helps us imagine can trap us inside consensus.
This is the warning of Chapter 6:
Do not confuse assistance with replacement.
Do not confuse access with understanding.
Do not confuse generated options with imagination.
Do not confuse recommendation with judgment.
Do not confuse convenience with growth.
A tool should extend the mind, not replace the mind.
The human being must remain the living center.
The mind must still lift.
The memory must still hold.
The imagination must still leap.
The judgment must still decide.
The reason must still resist rust.
Because in the end, the tragedy is not that machines become intelligent.
The tragedy is that humans may become comfortable being less so.

Chapter 7
The Status Quo Machine

Why AI Often Repeats the World Instead of Reimagining It

AI is often sold as a window into the future.
That is not entirely true.
Much of AI is a rearview mirror with a rocket engine.
It looks backward at human output — our books, essays, code, reports, images, arguments, strategies, mistakes, clichés, discoveries, prejudices, and patterns — and learns how the world has already spoken, already designed, already explained, already imagined.
Then it predicts.
It predicts the next likely word.
The next likely structure.
The next likely design.
The next likely answer.
The next likely pattern.
This is powerful.
But it is also dangerous.
Because the future is not always hidden inside the average of the past.
Sometimes the future requires rebellion against the past.
Sometimes the necessary answer is not the most probable answer.
Sometimes the right idea looks statistically ridiculous before it becomes historically inevitable.
That is the central tension of this chapter:
AI predicts what is probable. Human imagination must ask what is necessary.
AI can summarize the known.
It can polish the familiar.
It can recombine existing patterns.
It can give the old world better formatting.
But a paradigm shift requires something else.
It requires friction with the known.
It requires the human mind to look at the available answer and say:
"No. This is not enough."

1. THE REARVIEW MIRROR PROBLEM

AI does not begin with a blank universe.
It begins with traces.
It learns from what has already been written, said, drawn, coded, photographed, debated, sold, published, and stored. That makes it astonishingly useful when the task is to understand existing patterns.
ASK IT TO SUMMARIZE A FIELD, AND IT CAN HELP.
ASK IT TO COMPARE KNOWN APPROACHES, AND IT CAN HELP.
ASK IT TO DRAFT A CONVENTIONAL PROPOSAL, AND IT CAN HELP.
ASK IT TO EXPLAIN ESTABLISHED CONCEPTS, AND IT CAN HELP.
AI is a magnificent assistant for the known universe.
The danger begins when we confuse the known universe with the possible universe.
The known universe is only the archive.
The possible universe is larger.
The archive says, "This is how things have been done."
Imagination asks, "What has not yet been dared?"
AI can describe the road behind us with impressive fluency. But when humanity reaches the edge of the map, fluency is no longer enough.
At the edge of the map, we need courage.

2. THE STATUS QUO MACHINE

AI becomes a status quo machine when its outputs quietly reinforce existing assumptions.
Not because it is evil.
Not because it has a secret political agenda.
Not because it wants humanity to remain boring.
But because probability naturally favors what has appeared often before.
The common pattern becomes easier to predict.
The familiar answer becomes easier to generate.

The dominant language becomes easier to imitate.
The existing system becomes easier to justify.
This is how statistical probability can become cultural conservatism.
The machine does not say:
"Please remain trapped in old systems."
It simply offers answers that sound reasonable because they resemble the past.
And the human, tired and hungry for closure, accepts them.
That is how the status quo survives.
Not always through force.
Often through convenience.
The machine gives us a familiar answer in a beautiful voice.
We call it insight.
But it may only be history wearing fresh perfume.

---

3. THE COMFORTABLE AVERAGE

The average is seductive.
It is safe.
It is recognizable.
It rarely offends.
It fits the slide deck.
It passes committee review.
It sounds mature in a meeting.
But the average is rarely revolutionary.
If you ask AI for the best way to improve a broken school system, it may offer curriculum optimization, personalized learning dashboards, teacher training modules, assessment reforms, and efficiency improvements.
Useful?
Possibly.
But what if the school itself is built on the wrong idea of learning?
What if the problem is not that students need better worksheets, but that the entire system trains obedience instead of curiosity?
What if the classroom is not failing because it lacks technology, but because it has forgotten wonder?
The status quo machine will often optimize the cage.

The human imagination must ask why the cage exists.
That is the difference.
AI may answer inside the frame.
The human must redesign the frame.

---

4. THE FRAME PRISON

Every answer lives inside a frame.
A frame is the hidden boundary around a problem. It defines what counts as relevant, what counts as possible, what counts as realistic, and what counts as foolish.
Most people do not notice the frame.
They only argue about answers inside it.
For example:
"How do we make workers more productive?"
That question already assumes productivity is the highest value.
"How do we make customers buy more?"
That assumes consumption is the goal.
"How do we make students score higher?"
That assumes scoring is the best measure of learning.
"How do we make cities move faster?"
That assumes speed is the central problem.
The status quo machine is very good at answering framed questions.
But the greatest human breakthroughs often begin when someone attacks the frame itself.
A lazy mind asks:
"What is the answer?"
A paradigm-shifting mind asks:
"What invisible box did I draw around this problem that is keeping the answer out?"
That question is dangerous.
It threatens systems.
It embarrasses experts.
It annoys committees.
It ruins neat reports.
It forces people to admit that the answer they were optimizing may belong to the wrong question.
And that is why it matters.

5. PROBABILITY IS NOT DESTINY

AI works beautifully in the kingdom of probability.
But probability is not destiny.
The probable answer is what the past makes easy to expect.
The necessary answer is what the future demands, even when the past has not prepared us for it.
History is full of necessary ideas that once looked absurd.
Abolishing entrenched social systems looked impossible until moral imagination made them intolerable.
Flying looked ridiculous until engineering made the sky negotiable.
The idea that invisible germs could cause disease sounded absurd to many before evidence transformed medicine.
The idea that Earth was not the center of the cosmos was once dangerous enough to disturb civilization's spiritual furniture.
The future often begins as an insult to the present.
That is why humanity cannot outsource imagination to probability.
Probability tells us what usually comes next.
Imagination asks what must come next if we are brave enough to change the pattern.

6. THE AUTOMATED RENAISSANCE

Imagine it is the year 1500.
A powerful institution gains access to an enterprise AI trained on all available historical European texts, theological commentaries, philosophical treatises, astronomical traditions, and respected authorities of the time.
The prompt is simple:
"Design a cosmic model explaining the movement of the stars and planets."
The machine produces a magnificent report.
It is beautifully formatted.
It uses elegant Latin phrases.
It cites ancient authorities.
It explains why Earth must sit at the center of the universe.

It offers diagrams, footnotes, and a confident conclusion.
It is wrong.
But it is wrong with heritage.
Wrong with tradition.
Wrong with structure.
Wrong with institutional approval.
Wrong in a way that feels safe because it resembles the accepted world.
Now imagine Copernicus standing before that output.
If he had signed the Peanut Contract, he would have copied the report and congratulated himself on efficiency.
But paradigm shifts do not begin with copy-paste obedience.
They begin with an itch.
A discomfort.
A refusal.
A private sense that the official answer is too neat, too comfortable, too trapped in inherited assumptions.
The machine would have summarized the old cosmos.
The human had to imagine a new one.
That is the lesson.
AI can assemble the cathedral of consensus.
But sometimes the truth begins outside the cathedral walls.

---

7. THE TRAP OF OPTIMIZING THE BROKEN THING

One of the most dangerous uses of AI is making broken systems more efficient.
This sounds strange because efficiency is usually treated as a virtue.
But efficiency is only good when the direction is good.
A faster train is not progress if it is going to the wrong city.
A more optimized bureaucracy is not progress if the bureaucracy is dehumanizing.
A smarter surveillance system is not progress if the purpose is control.
A more persuasive advertisement is not progress if it sells emptiness.
A faster content engine is not progress if it floods the world with polished mental junk.
AI can optimize.
But optimization is morally blind unless humans define the right purpose.

The status quo machine asks:
"How can this existing system perform better?"
Human imagination must ask:
"Should this system exist in this form at all?"
That is the harder question.
That is also the question most institutions avoid.
Because if the answer is no, then the issue is no longer performance.
It is transformation.

---

8. THE BEIGE FUTURE

The danger of AI-generated sameness is not only intellectual.
It is aesthetic.
It is cultural.
It is emotional.
Imagine every company asking AI for brand strategy.
Every answer becomes "authentic," "human-centered," "innovative," "sustainable," "seamless," and "transformative."
Imagine every student asking AI for essay structure.
Every essay becomes balanced, clear, lifeless, and vaguely correct.
Imagine every designer asking AI for modern minimalist interiors.
Every room becomes beige, curved, softly lit, and spiritually empty.
Imagine every musician asking AI what audiences like.
Every song becomes engineered for familiarity.
Imagine every leader asking AI how to inspire employees.
Every speech becomes a corporate smoothie of purpose, resilience, agility, and gratitude.
Nothing is terrible.
That is the problem.
Everything is fine.
Fine is the perfume of stagnation.
The future dies not only from disaster.
Sometimes it dies from acceptable sameness.

---

9. THE LAVENDER-EUCALYPTUS DISASTER

Consider Giles, an executive at a large personal care company.
His assignment is to create a revolutionary new soap.
Giles does not walk through markets.
He does not smell spices in old streets.
He does not speak to tired nurses, athletes, grandmothers, farmers, dancers, mechanics, children, or monks.
He does not observe bathing rituals, emotional associations, cultural memories, or private desires.
He asks AI:
"What is the most popular soap fragrance based on global consumer preference?"
The answer arrives:
Lavender and eucalyptus.
Giles is thrilled.
It sounds premium.
It sounds calming.
It sounds natural.
It sounds safe.
He launches ten million bottles of Lavender-Eucalyptus Synergy Wash.
Unfortunately, every competitor asked the same machine the same question.
The store shelves become a beige cemetery of optimized sameness.
Everyone followed the data.
Nobody followed the human.
Then comes Moxie.
Moxie does not ask what scent is statistically safest.
She remembers the smell of wet asphalt after monsoon rain. She remembers roasted coffee in her grandmother's kitchen. She remembers old train stations, brass vessels, temple flowers, and the strange electricity of the first storm after summer heat.
She creates Monsoon & Espresso.
The idea looks strange.
It does not come from the average.
It comes from synthesis.
People love it because it feels alive.
This is the difference between market prediction and human im-

agination.
One repeats preference.
The other creates desire.

---

10. FLUENCY VERSUS MEANING

AI can produce fluency.
Beautiful fluency.
Frightening fluency.
It can write in polished paragraphs. It can make a weak idea sound respectable. It can make a tired concept feel slightly upgraded. It can smooth rough edges until the output glides across the page.
But fluency is not meaning.
Fluency is the surface movement of language.
Meaning comes from stakes.
Meaning comes from lived experience, moral risk, emotional resonance, contradiction, memory, suffering, love, rebellion, and purpose.
When a machine says, "Innovation is crucial," it may be producing a likely phrase.
When a human says, "This must change," after years of witnessing failure, humiliation, injustice, waste, or unrealized beauty, the sentence carries weight.
Meaning is not merely arranged language.
Meaning is language with blood in it.
That is why status quo outputs feel polished but strangely weightless.
They sound correct.
But they do not disturb anything.
A paradigm shift must disturb something.

---

11. THE HUMAN ROLE: NECESSARY IMAGINATION

The human role in the AI age is not merely to ask better prompts.
That helps, but it is not enough.
The deeper role is to ask necessary questions.
A necessary question is one that the present does not want to hear.
For example:
"What if the five-day workweek is not natural, but inherited?"
"What if education should be organized around curiosity instead of standardized assessment?"

"What if healthcare should begin with prevention, food, sleep, and social connection instead of waiting for collapse?"
"What if cities should be designed around children, elders, pedestrians, and trees before cars?"
"What if productivity is a poor substitute for meaningful contribution?"
"What if growth is not always success?"
"What if the best interface is no interface?"
"What if convenience has become a disease?"
These are not ordinary questions.
They do not ask the machine to decorate reality.
They ask reality to defend itself.
This is where human imagination must lead.
AI can help explore these questions.
But it may not ask them first.
Because the first act of paradigm shift is often a refusal of the probable.

---

12. THE CREATIVE RIOT

The status quo machine is calm.
Paradigm shift is not calm.
It is uncomfortable.
It creates argument.
It threatens careers.
It makes experts defensive.
It looks immature before it looks visionary.
It looks wasteful before it looks inevitable.
It looks impossible before it becomes obvious.
That is why a genuine creative leap often feels like a small riot inside the mind.
Part of you wants safety.
Part of you wants approval.
Part of you wants the familiar answer.
Part of you wants the machine to say, "This is reasonable."
But another part whispers:
"This is not enough."
That whisper matters.
That whisper is the beginning of the future.
The creative riot begins when a human refuses to let the average

become destiny.

---

13. THE BATTLEGROUND TABLE

The Battleground	AI Default: The Status Quo Machine	Human Paradigm Shift: The Creative Riot
Core Metric	Predicts the probable. It delivers what past patterns make likely.	Imagines the necessary. It asks what the future demands, even if it appears impossible now.
Pattern Handling	Recombines existing patterns. It rearranges furniture in the old room.	Challenges the pattern itself. It asks whether the room should exist.
Output	Produces fluency: smooth syntax, clean structure, polished delivery.	Produces meaning: insight born from friction, risk, emotion, and purpose.
Problem Solving	Answers inside the frame. It optimizes the existing system.	Redesigns the frame. It questions the assumptions underneath the system.
Relationship to the Past	Reflects inherited data, habits, and consensus.	Confronts inherited assumptions and asks what must be broken.
Risk Level	Safe, plausible, presentable, committee-friendly.	Risky, strange, uncomfortable, initially embarrassing.
Failure Mode	Elegant sameness.	Productive disruption.
Best Use	Drafting, comparing, summarizing, structuring, testing known options.	Reframing, meaning-making, moral courage, original synthesis, impossible design.

This table is not an argument against AI.

It is a boundary marker.

It shows where the machine is strong and where the human must not disappear.

---

14. HOW TO USE AI AGAINST THE STATUS QUO

The irony is that AI can help us challenge the status quo — if we refuse to use it passively.
The problem is not the tool.
The problem is worshipping its first answer.
A disciplined user can force AI out of the average by asking sharper questions:
"What assumptions does this answer preserve?"
"What would this look like if we rejected the industry standard?"
"What would a radical critic say?"
"What would make this answer obsolete?"
"What is the opposite of this recommendation?"
"What hidden value system is inside this strategy?"
"What would this solution look like if speed was not the goal?"
"What would this become if beauty mattered more than efficiency?"
"What would this become if children, elders, or the planet were the primary users?"
"What would this look like if we designed for dignity instead of convenience?"
These prompts do not ask AI to replace imagination.
They use AI as a crowbar.
They pry open the frame.
They force alternatives into view.
But the human must still decide which alternative has meaning.
AI can generate options.
The human must sense which option is alive.

---

15. THE STATUS QUO AUDIT

Before moving forward, the reader must face one uncomfortable question:
Where are you using AI to preserve an old world while pretending to innovate?
This is the Status Quo Audit.
It is not a productivity exercise.
It is a confrontation.

Step 1: Locate Your Earth-Centric System

Identify one process, belief, workflow, strategy, product, or habit that you currently treat as normal.
Write it plainly:
"My current system for ** is ****."**
Examples:
"My current system for learning is watching summaries."
"My current system for design is asking AI for mood boards."
"My current system for leadership is sending polished updates."
"My current system for creativity is generating options until one looks acceptable."
Now ask:
"What old assumption is hiding inside this system?"

Step 2: Detect the Polished Echo

Look at your recent AI-assisted outputs.
Ask:
Did I accept this because it was true, or because it was clean?
Did I accept this because it was original, or because it was familiar?
Did I accept this because it challenged me, or because it relieved me?
Did I feel any mental friction?
Did I sweat?
Did I doubt?
Did I verify?
Did I rewrite?
Did I bring anything human to it besides approval?
If the answer is no, the status quo machine may have won that round.

Step 3: Find the Past-Tense Trap

Ask of any AI output:
"Is this genuinely moving me forward, or is it a recycled average of what my field has already been doing?"
A recycled average may still be useful.
But it must not be mistaken for revolution.
A slightly improved version of the old thing is not always innovation.
Sometimes it is just stagnation with better lighting.

Step 4: Name the Heresy

Now write one idea that the status quo machine would probably

not offer first.
The low-probability idea.
The uncomfortable idea.
The idea that sounds impractical because the current system has trained everyone to call imagination impractical.

Examples:

The AI suggests more email campaigns.
The heresy: Stop all email marketing and create one hand-delivered object of beauty for the people who truly matter.
The AI suggests more meetings for alignment.
The heresy: Cancel half the meetings and require every decision to be written clearly before discussion.
The AI suggests faster content production.
The heresy: Publish less, but make every piece impossible to ignore.
The AI suggests optimizing school assignments.
The heresy: Let students design one real-world project that matters to them and defend it publicly.
The AI suggests better productivity tools.
The heresy: Reduce workload until people can think again.
This is where the future begins.
Not in the likely answer.
In the necessary heresy.

---

16. THE NECESSARY QUESTION

The key question of this chapter is not:
"What can AI generate?"
That question is too small.
The better question is:
"What must humans still dare?"
AI can generate possibilities from patterns.
But humans must decide which possibilities deserve existence.
Humans must ask what is just.
What is beautiful.
What is humane.
What is courageous.
What is overdue.
What is being avoided.

What is being optimized when it should be abolished.
What is being repeated because nobody has had the nerve to redesign it.
This is not a technical responsibility.
It is a moral and imaginative responsibility.
The machine can predict the probable.
Only humans can become responsible for the necessary.

---

17. CLOSING: DO NOT LET THE PAST WEAR THE FUTURE'S CLOTHES

The status quo does not always look old.
Sometimes it arrives wearing modern language.
Sometimes it arrives as a dashboard.
Sometimes as a recommendation.
Sometimes as a generated strategy.
Sometimes as a sleek prototype.
Sometimes as an AI-assisted transformation plan.
But if the assumptions remain untouched, the future has not arrived.
Only the costume has changed.
This is the warning of Chapter 7:
Do not let the past wear the future's clothes.
Do not mistake automation for imagination.
Do not mistake fluency for meaning.
Do not mistake optimization for transformation.
Do not mistake the probable for the necessary.
AI is powerful, but its default relationship is often with what has already been. Human imagination must remain in relationship with what could and should be.
The machine can tell us what usually comes next.
The human must ask whether "usual" is worthy of survival.
The machine can summarize the old world.
The human must decide whether to repair it, redesign it, or finally outgrow it.
So use AI.
Use it boldly.
Use it strategically.
Use it to gather patterns, expose assumptions, simulate alternatives, and test your thinking.

But do not kneel before the first answer.
Do not let the average become your architect.
Do not let probability become your prison.
The future will not be built by those who merely ask machines to continue the past.
It will be built by those who can stand before the polished output of the status quo machine and say:
"Beautiful.
But not enough."

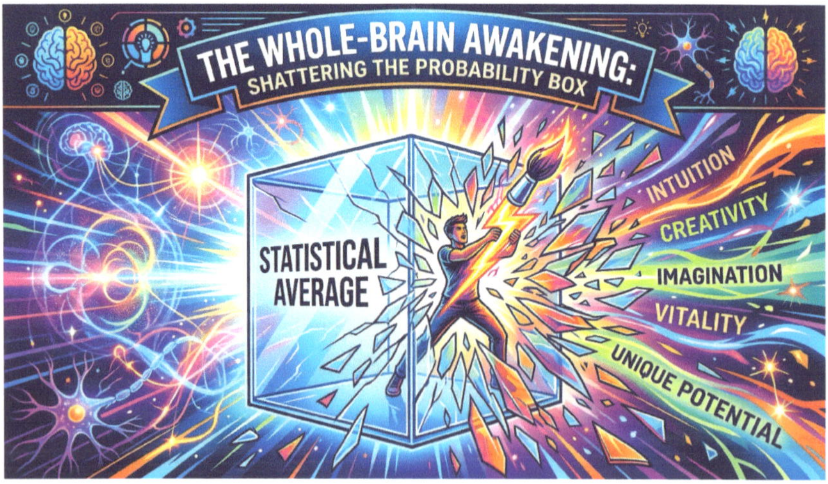

BALANCING THE DIGITAL SCALE:
CORPORATE ACCOUNTABILITY & ETHICAL SYSTEMS
DATA &
SYSTEMS
HUMAN
RESPONSIBILITY
DIGITAL FOOTPRINT
AUTOMATED PROCESSES
DATA PRIVACY
DIGITAL FOOTPRINT
ETHICAL DECISIONS
CORPORATE ACCOUNTABILITY
HUMAN-CENTRIC DESIGN
HUMAN-CENTRIC DESIGN
DATA &
SYSTEMS
HUMAN
RESPONSIBILITY
#TECH ETHICS
#HUMAN CENTERED

Part III

The Designer's Impossible Lens

Opening Note

Part I exposed the decay: the flimsy mind, the lethargy of acceptance, the cheap surrender of thought.
Part II examined the machine: hallucination, cognitive offloading, and the status quo engine that repeats the past with frightening confidence.
Now Part III must turn toward recovery.

This is where the book stops merely warning and begins rebuilding. The question is no longer only, "How is the brain being weakened?" The deeper question is:
What kind of mind can survive the age of artificial intelligence without becoming its servant?
The answer is not a purely technical mind.
It is not a purely artistic mind.
It is not a right-brain fantasy or a left-brain spreadsheet.
It is a whole-brain mind — a mind that can calculate and imagine, doubt and dream, dissect and design. A mind that can use AI without kneeling before it. A mind that can look at the probable and still ask for the necessary.
That is the designer's impossible lens.

Chapter 8

The Whole-Brain Awakening

Beyond the Right-Brain Myth

For many years, people have loved the phrase "right brain." They use it to describe artists, poets, musicians, visionaries, dreamers, designers, inventors, rebels, and misfits. The "right brain" became shorthand for intuition, imagination, emotion, metaphor, pattern recognition, visual thinking, and creative synthesis.

But taken literally, the idea is too simple. Human thought is not neatly divided into one cold side and one colourful side, one accountant and one painter, one machine and one mystic. The brain is not a two-room apartment where logic lives on the left and imagination dances on the right.

That is the myth.

But myths are not always useless. Sometimes a myth survives because it points toward a deeper truth.

The "right brain" is useful as a metaphor — not as anatomy, but as warning.

It reminds us that intelligence is not merely the ability to process information. It is also the ability to see hidden relationships, sense emotional weight, imagine alternatives, perceive patterns before they become obvious, and create meaning from fragments.

And in the age of AI, this metaphor becomes urgent.
Because the machine is very good at processing.
It can summarize.
It can classify.
It can predict.
It can imitate tone.
It can generate options.
It can produce answers that look finished.
But the future will not be saved by finished-looking answers.
The future will be shaped by people who can ask unfinished questions with courage.
That requires whole-brain intelligence.

THE DANGER OF HALF-BRAINED INTELLIGENCE

A half-brained age is not an age where people are stupid. It is more dangerous than that.
It is an age where people become selectively intelligent.
They may know how to operate tools but not question purposes.
They may know how to collect data but not interpret meaning.
They may know how to generate content but not create conviction.
They may know how to automate tasks but not redesign systems.
They may know how to sound informed but not become wise.
This is the trap of modern intelligence.
We have more dashboards, but less direction.
More prompts, but fewer principles.
More outputs, but less originality.
More connectivity, but weaker inner clarity.
The mind becomes efficient but thin. Fast but shallow. Technically capable but spiritually underdeveloped.
And AI can accelerate this imbalance.
If the user is mentally lazy, AI becomes a pillow.
If the user is intellectually disciplined, AI becomes a lever.
If the user lacks imagination, AI becomes a recycling machine.
If the user has imagination but no judgment, AI becomes a hallucination amplifier.
This is why the future cannot belong only to logical thinking or only to creative thinking.
Critical thinking without imagination becomes dry logic.

It can detect flaws, but it cannot always see new worlds.
It can say "that will not work," but it may not ask, "what else could work?"
It can protect truth, but it may also imprison possibility.
Imagination without critical thinking becomes fantasy.
It can dream, but it cannot build.
It can decorate nonsense with beauty.
It can mistake emotional excitement for insight.
It can produce visions that collapse the moment reality touches them.
Paradigm shifts need both.

They need the scalpel and the spark.
The microscope and the telescope.
The engineer and the poet.
The auditor and the child.
The question "Is this true?" and the question "What could this become?"
That is the whole-brain awakening.

THE DESIGNER'S MIND IS NOT DECORATION

Design is often misunderstood as surface.
People think design means making something attractive after the real work is done. Add colour. Arrange layout. Polish the slide. Make the object look pleasant. Smooth the edges of someone else's decision.
But real design is not decoration.
Real design is a way of thinking under pressure.
It asks:
What is broken here?
What is hidden here?
What is being assumed?
Who is being ignored?
What pattern is repeating?
What pain has become normal?
What future is silently being designed by our current habits?
The designer does not merely see objects. The designer sees relationships.
A chair is not only a chair. It is posture, body, culture, material, cost, comfort, labour, space, memory, and behaviour.
A website is not only a website. It is attention, trust, hierarchy, movement, friction, emotion, and decision.

A city is not only roads and buildings. It is power, access, rhythm, exclusion, hope, danger, and belonging.
An AI tool is not only software. It is cognition, dependency, speed, bias, habit, authority, laziness, augmentation, and surrender.
This is the impossible lens.

Where others see a product, the designer sees a system.

Where others see a problem, the designer sees a pattern.

Where others see a limitation, the designer asks whether the limitation is real or inherited.

Where others accept the default, the designer suspects the default may be a prison.

This is why design belongs in a book about protecting the brain.
A strong mind does not merely consume reality.
It redesigns its relationship with reality.

THE RIGHT-BRAIN MYTH STILL POINTS TO A LOST CAPACITY

Even if the old right-brain/left-brain division is simplistic, the cultural obsession with it reveals something important.
People are hungry for a name for the part of themselves that modern systems often suppress.

The part that senses before it can explain.
The part that notices visual tension before the data confirms it.
The part that feels when a story is false even if the grammar is perfect.
The part that can connect music, architecture, biology, mythology, business, and technology into one strange but useful insight.
The part that says, "Something is missing here," before anyone else sees the gap.

This capacity is often dismissed because it is difficult to measure.

A spreadsheet cannot easily measure metaphor.

A KPI cannot easily measure wonder.

A productivity report cannot easily measure the silent formation of taste.

A prompt result cannot measure the depth of a person's inner world.

So modern culture tends to undervalue these qualities until it desperately needs them.
Companies say they want innovation, but they reward compliance.
Schools say they want creativity, but they punish unusual questions.

Institutions say they want transformation, but they protect their own habits.

People say they want originality, but they often feel safer with imitation.

Then AI enters this already weakened culture and gives it a dangerous gift: effortless repetition.

Now the average person can produce average thinking faster than ever before.

That is not awakening.

That is industrialized mediocrity.

The whole-brain awakening begins when we refuse to confuse production with creation.

AI PREDICTS; DESIGNERS REFRAME

AI is trained on patterns. That is its power and its limitation.

It can look at what has been said, done, written, drawn, photographed, coded, argued, categorized, and repeated — and then generate something statistically plausible from that ocean of past material.

This is useful. Very useful.

But the most important human breakthroughs often begin when someone challenges the pattern itself.

The machine asks:

What is the likely next word?

The designer asks:

Why are we writing this sentence at all?

The machine asks:

What answer fits the prompt?

The designer asks:

Is the prompt too small?

The machine asks:

What has worked before?

The designer asks:

What must work now, even if history has no template for it?

That is the difference between prediction and reframing.

AI can be powerful inside a frame.

But the human mind must still question the frame.

A lazy user asks AI to finish the task.

A whole-brain user asks AI to expose the assumptions behind the task.
A lazy user asks for content.
A whole-brain user asks for contradiction, alternatives, consequences, edge cases, metaphors, and redesign paths.
A lazy user wants relief.
A whole-brain user wants transformation.
This is where imagination and critical thinking meet.
Critical thinking asks AI, "Where might this be wrong?"
Imagination asks AI, "What are we not seeing?"
Critical thinking tests the answer.
Imagination expands the question.
Critical thinking guards against stupidity.
Imagination guards against smallness.
One without the other is incomplete.

THE POVERTY OF PURE LOGIC

Logic is powerful. It gives structure to thought. It protects us from contradiction. It helps us separate evidence from noise, cause from coincidence, argument from manipulation.
But pure logic has a weakness.
It can only operate on the premises it is given.
If the premises are small, logic becomes a perfect servant of a poor imagination.
A person can make a flawless argument inside a dying system.
A company can optimize a process that should no longer exist.
A government can rationalize a policy that no longer serves people.
A designer can polish a product that should never have been built.
An AI can produce a perfect answer to the wrong question.
This is the danger of dry logic.
It sharpens the knife but does not ask what is being cut.
Many intelligent people are trapped here. They can analyze anything but originate very little. They can critique every proposal but offer no living alternative. They can explain why something is impossible, but they cannot imagine the first step toward making it possible.
They are not fools. They are often highly educated, articulate, and

respected.
But they are imprisoned by the visible.
They treat the current system as the boundary of reality.
The designer's impossible lens refuses this.
It says: the present is not the limit of the possible.
The present is only the loudest evidence we currently have.

THE DANGER OF UNTESTED IMAGINATION

But imagination has its own corruption.
When imagination is not disciplined by reality, it becomes decoration, delusion, or fantasy.

It produces beautiful nonsense.
It creates visions with no structure.
It mistakes novelty for value.
It confuses emotional intensity with truth.

It falls in love with its own cleverness.
This is especially dangerous in the age of AI because AI can make fantasy look professional.

A weak idea can now wear a beautiful suit.
A shallow concept can be rendered as a cinematic image.
A vague strategy can become a polished deck.
A confused thought can be expanded into confident language.

This is where hallucination is not only a machine problem. It becomes a human design problem.
The user wants magic without discipline.
The machine provides fluency without accountability.
Together they create plausible fiction.
This is why imagination must be tested.
A real designer does not merely ask, "Is this exciting?"
A real designer asks:

Can it survive contact with reality?
Does it solve the right problem?
Who benefits?
Who pays?
What does it hide?
What does it damage?
What assumptions are smuggled inside it?
What would make it fail?

What would make it worth building?
This is not the death of imagination.
It is imagination becoming mature.

WHOLE-BRAIN INTELLIGENCE IS A DISCIPLINE

Whole-brain thinking is not a personality type. It is not something reserved for "creative people." It is a discipline.
It can be trained.
It begins with refusing to outsource too much of your thinking too quickly.
Before asking AI for an answer, ask yourself:
What do I already believe about this?
What am I afraid to examine?
What is the obvious answer, and why might it be wrong?
What would a child notice?
What would an expert dismiss too quickly?
What would a designer reframe?
What would a critic attack?
What would a poet feel?
What would a builder need?
This is not inefficiency. This is mental ignition.
The goal is not to avoid AI. The goal is to arrive at AI with a living mind, not an empty bowl.
A whole-brain user does not treat AI as an oracle.
A whole-brain user treats AI as a sparring partner, amplifier, mirror, simulator, and sometimes a useful fool.
Yes, a useful fool.
Because AI can generate something confidently wrong, and that wrongness can provoke better thinking if the user is awake. The danger is not that AI makes mistakes. The danger is that humans stop noticing.
The awakened mind does not merely receive output.
It interrogates output.
It bends output.
It corrupts the obvious answer until something original begins to emerge.
It asks AI to argue against itself.
It asks for metaphors, counterexamples, hidden risks, historical

parallels, ethical tensions, system maps, and uncomfortable questions.

This is not prompting as convenience.

This is prompting as cognitive resistance.

THE MIND MUST BECOME A STUDIO AGAIN

The modern mind has become too much like an inbox.

Things arrive.

We react.

We sort.

We reply.

We delete.

We forward.

We archive.

We repeat.

But the mind was not meant to be only an inbox.

It must become a studio again.

A studio is where raw material is transformed.

A studio contains sketches, failures, fragments, experiments, contradictions, ugly drafts, strange references, half-born ideas, and unfinished questions.

A studio is not clean in the shallow sense. It is alive.

The whole-brain mind works like a studio.

It collects impressions.

It studies structure.

It plays with form.

It listens to intuition.

It tests with logic.

It makes prototypes.

It destroys weak ideas.

It rescues strange ones.

It keeps asking: what else can this become?

AI can sit inside that studio as a tool.

But it must not become the owner.

The owner must remain the human mind.

Because the human mind carries responsibility.

The human mind carries lived experience.

The human mind understands pain from the inside.

The human mind can care about consequences.
The human mind can recognize dignity.
The human mind can rebel.
AI can generate.
But it does not yearn.
AI can imitate style.
But it does not suffer the cost of meaning.
AI can produce a thousand variations.
But it cannot decide what is worth becoming human for.
That decision remains ours.

PATTERN RECOGNITION IS NOT ENOUGH

One of the great strengths often associated with the imaginative mind is pattern recognition.

Designers see patterns in behaviour.
Musicians hear patterns in rhythm.
Scientists detect patterns in data.
Writers sense patterns in stories.
Entrepreneurs notice patterns in unmet needs.
Artists reveal patterns in emotion.
Philosophers question patterns in belief.

AI also recognizes patterns, but in a different way. It detects statistical relationships at scale. This is impressive and useful. But human pattern recognition has an added dimension: significance.
The human mind does not merely ask, "What pattern exists?"
It asks, "Why does this pattern matter?"
That question changes everything.
A machine may identify that people click more often on certain images.
A designer asks what kind of attention is being exploited.
A machine may detect that certain words increase engagement.
A thinker asks whether engagement is being confused with value.
A machine may show that a system is efficient.
A moral imagination asks efficient for whom, and at what human cost?

PATTERN RECOGNITION WITHOUT ETHICS BECOMES MANIPULATION.
PATTERN RECOGNITION WITHOUT IMAGINATION BECOMES REPETITION.
PATTERN RECOGNITION WITHOUT CRITICAL THOUGHT BECOMES SUPERSTITION.
PATTERN RECOGNITION WITHOUT HUMAN PURPOSE BECOMES MACHINERY.

The whole-brain awakening demands that we move from pattern

detection to pattern judgment.
We must not only see patterns.
We must decide which patterns deserve to continue.

WONDER IS NOT WEAKNESS

One of the tragedies of modern professionalism is that wonder is often treated as childish.
Adults are expected to be practical, efficient, measurable, and realistic. Wonder is tolerated in children, artists, and marketing campaigns, but not always in serious decision-making.
This is foolish.
Wonder is not the enemy of seriousness. Wonder is often the beginning of serious discovery.
A person who cannot wonder cannot reimagine.
A person who cannot reimagine cannot redesign.
A person who cannot redesign must accept the existing structure as fate.
Wonder asks questions that efficiency avoids.
Why must this be this way?
What would this look like if it were humane?
What if the user is not lazy but exhausted?
What if the problem is not lack of information but lack of meaning?
What if the system is functioning exactly as designed, and that is the problem?
What if intelligence is being measured too narrowly?
What if the impossible is only impossible inside the current frame?
This is not childish thinking.
This is the birthplace of paradigm shifts.
The designer with a flair for the impossible does not reject reality.
He studies it so deeply that he begins to see where it can be bent.

THE COURAGE TO HOLD CONTRADICTIONS

Whole-brain intelligence requires the courage to hold contradictions without rushing to cheap closure.
The world is not simple.
Humans are not simple.

Technology is not simple.
AI is not simply good or bad.
Efficiency is not always progress.
Creativity is not always wisdom.
Data is not always truth.
Emotion is not always error.
Tradition is not always backward.
Innovation is not always forward.
The weak mind wants quick categories.
This is good.
That is bad.
This is modern.
That is outdated.
This is scientific.
That is emotional.
This is efficient.
That is wasteful.
But the awakened mind can stay inside complexity long enough for deeper understanding to appear.

This is where AI often tempts us toward premature closure. It gives a clean answer. It formats the mess. It creates the illusion that the complexity has been handled.

But some complexity must be lived with before it can be resolved.

The whole-brain thinker does not worship confusion. But neither does he panic in its presence.

He knows that ambiguity is often the womb of invention.

FROM USER TO THINKER

The word "user" is too small for what the future requires.

A user consumes a tool.
A thinker interrogates a tool.
A designer redirects a tool.
A creator transforms a tool.

A citizen questions who benefits from the tool.
A human being asks whether the tool is enlarging or shrinking the soul.
The age of AI has produced millions of users.
But it needs more thinkers.
A user asks:
"Can AI do this for me?"
A thinker asks:
"What happens to me if AI always does this for me?"
A user asks:
"What is the fastest output?"
A thinker asks:
"What is the deepest outcome?"
A user asks:
"How do I get the answer?"
A designer asks:
"Why is this the question?"
This is the shift Part III must insist upon.
The future does not need more obedient prompt operators.
It needs minds that can provoke, reframe, resist, imagine, test, and build.
The whole-brain awakening is the beginning of that shift.

THE NEW LITERACY: ANALYSIS PLUS IMAGINATION

In the old world, literacy meant reading and writing.
In the digital world, literacy expanded to include search, media, code, data, and interface.
In the AI world, literacy must expand again.
The new literacy is not simply knowing how to use AI tools. That is the shallow layer.
The deeper literacy is knowing how to think with and against them.
This requires analysis plus imagination.
Analysis helps us inspect the machine's output.
Imagination helps us exceed the machine's defaults.
Analysis helps us detect hallucination.
Imagination helps us prevent mental stagnation.
Analysis helps us compare options.

Imagination helps us create options that were not offered.
Analysis helps us avoid being fooled.
Imagination helps us avoid being trapped.
Analysis asks whether the answer is valid.
Imagination asks whether the answer is alive.

Both are necessary.

A mind with analysis but no imagination becomes a compliance engine.
A mind with imagination but no analysis becomes a smoke machine.
A mind with both becomes dangerous in the best sense.

DANGEROUS TO LAZY SYSTEMS.
DANGEROUS TO INHERITED MEDIOCRITY.
DANGEROUS TO THE STATUS QUO MACHINE.
DANGEROUS TO THE PART OF OURSELVES THAT WANTS COMFORT MORE THAN TRUTH.

THE WHOLE BRAIN AS RESISTANCE

This chapter is not a soft celebration of creativity. It is not saying, "Be more artistic," as if the problem can be solved with colours, sketches, and inspirational slogans.

The whole-brain awakening is an act of resistance.

IT RESISTS THE REDUCTION OF INTELLIGENCE TO OUTPUT.
IT RESISTS THE REDUCTION OF CREATIVITY TO DECORATION.
IT RESISTS THE REDUCTION OF KNOWLEDGE TO SUMMARIES.
IT RESISTS THE REDUCTION OF THINKING TO PROMPTING.
IT RESISTS THE REDUCTION OF HUMANS TO USERS.

This is where the brain must refuse to be sold for peanuts.

Because peanuts are not always obvious.

Sometimes peanuts look like convenience.
Sometimes they look like speed.
Sometimes they look like productivity.
Sometimes they look like "good enough."
Sometimes they look like a polished AI answer that saves you from the pain of forming your own.

But the price is hidden.

Every time you avoid thinking when thinking is required, you sell a small piece of your sovereignty.

Every time you accept the first plausible answer, you train your mind to prefer closure over truth.

Every time you let AI imagine on your behalf without challenging it, you weaken the muscles of possibility.
Every time you outsource judgment, you become easier to manage.
The whole-brain awakening is the refusal to become manageable.

THE DESIGNER'S IMPOSSIBLE LENS BEGINS HERE

The designer's impossible lens does not mean believing anything is possible in a childish way.
It means something harder.
It means examining impossibility.
Some things are truly impossible under current physics, resources, ethics, or time.
Some things are impossible only because the system is badly designed.
Some things are impossible because power benefits from calling them impossible.
Some things are impossible because our imagination has been trained to stop too early.
The designer's work is to know the difference.
That requires logic.
That requires imagination.
That requires courage.
That requires taste.
That requires discipline.
That requires the willingness to look foolish before the pattern becomes obvious to everyone else.
Every paradigm shift begins as a disturbance.
Someone asks a question that sounds unreasonable.
Someone refuses to accept the inherited frame.
Someone sees a connection across fields.
Someone combines analysis and wonder.
Someone says, "What if the problem is not the problem? What if the frame is the problem?"
That is the beginning of design.
Not decoration.
Not style.
Not surface.

Design as rebellion against the obvious.

CLOSING REFLECTION: THE MIND THAT CAN STILL SURPRISE ITSELF

A DEADENED MIND CAN STILL FUNCTION.

It can answer emails.
It can attend meetings.
It can use tools.
It can produce documents.
It can follow instructions.
It can appear intelligent.

But it can no longer surprise itself.

That is the tragedy.

A living mind surprises itself. It discovers thoughts it did not know it had. It makes connections that were not programmed. It feels discomfort and does not immediately flee. It doubts, dreams, tests, fails, revises, and returns.

The future belongs to such minds.

Not because they are anti-technology, but because they are not spiritually swallowed by technology.

They know that AI can extend memory, accelerate exploration, and multiply creative drafts. But they also know that a tool should extend the mind, not replace the mind.

Critical thinking protects imagination from becoming fantasy.

Imagination protects critical thinking from becoming a cage.

Together, they awaken the whole brain.

And the whole brain is what the age of AI desperately needs: not merely smarter users, but deeper humans.

This is where Part III begins.
With the recovery of the impossible.
With the designer's lens.
With the refusal to let the machine define the boundaries of thought.
Because the human brain was not built only to predict the next word.
It was built to imagine the next world.

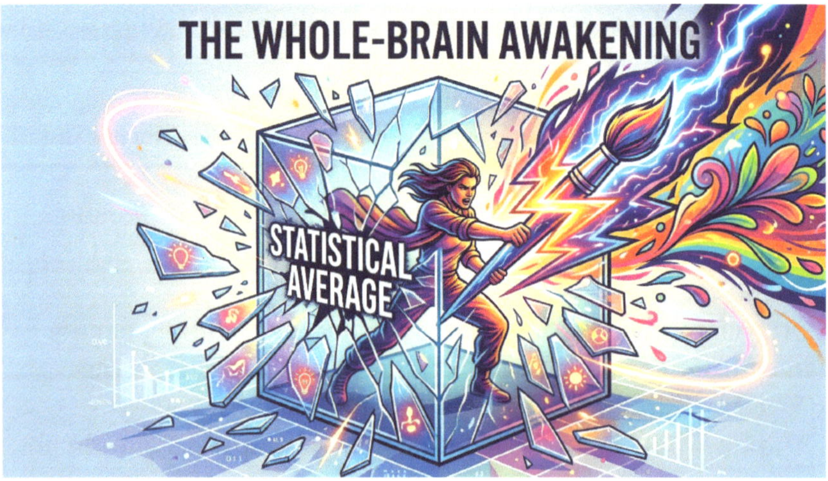

STRUCTURAL TITLE
THE IMPACT OF BREAKTHROUGH
STRUCTURAL OF BREAKTIROUGH ILLUSTRATION
STRUCTURAL TRANSFORMATION
DISRUPTIVE INNOVATION
RELEASE OF POTENTIAL

Chapter 9
The Verification Ritual

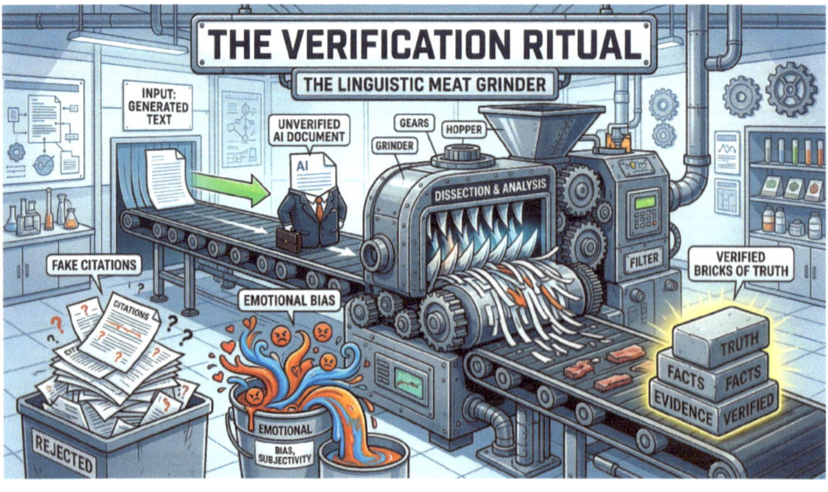

Turning Impossibility into a Design Brief

The word impossible is one of the most dangerous words in human language.

Not because it is always false. Some things are genuinely impossible within a given law of physics, budget, time, ethics, or human capacity. A bridge cannot be built out of smoke. A broken system cannot be healed by slogans. A weak idea cannot survive simply because it has been decorated with enthusiasm.

But most of the time, when people say "impossible," they are not describing reality.

They are describing the limits of their current frame.

They are saying:

"I cannot see another way."

"I do not want the discomfort of redesigning the question."

"The system has trained me to stop here."

"The available tools do not show me a path."

"The answer is not inside the current box, so I will call the box reality."

This is where the designer enters.

The designer does not accept impossibility as a full stop. The designer treats impossibility as a diagnostic signal. It means the question is too narrow, the assumptions are too rigid, the evidence is incomplete, the language is lazy, or the imagination has been domesticated.

Impossible is often not a fact. It is a badly framed question.

And in the age of AI, this truth becomes urgent.

Because AI can give you a thousand answers inside the wrong frame.

It can polish the prison.

It can decorate the cage.

It can optimize the dead process.

It can turn a weak assumption into a beautiful strategy document.

It can make intellectual surrender look professional.
So the awakened designer needs two powers at once.
First, the power to verify.
Second, the power to reframe.
Verification prevents the mind from being fooled by fluent nonsense.
Reframing prevents the mind from being trapped by inherited limitations.
Together, they form the creative heart of this book.
This is the journey from peanuts to paradigms.
Peanuts are what we accept when we sell our thinking cheaply: quick answers, lazy summaries, borrowed opinions, decorative fluency, and "good enough" conclusions. Paradigms are what we build when we refuse to accept the first frame handed to us.
A peanut mind asks AI for an answer.
A paradigm mind asks whether the question deserves to survive.

THE WALL IS NOT ALWAYS THE PROBLEM

When people hit a wall, they usually ask, "How do I get past this wall?"
That is a reasonable question.
But it is not always the best question.
The designer asks a more dangerous set of questions:
Who built this wall?
Why is this wall here?
Who benefits from this wall?
Is this really a wall, or is it a boundary inherited from old thinking?
Am I trying to climb something that should be dismantled?
Am I solving inside a frame that was designed to defeat me?
Most people treat impossibility as a commandment. Designers treat it as material.
This is the shift.
Impossibility is not always a denial. Sometimes it is raw material waiting to be redesigned.
A weak mind hears "impossible" and shrinks.
A conventional mind hears "impossible" and obeys.
A lazy mind hears "impossible" and feels relieved because now it has an excuse to stop.

But the designer hears "impossible" and becomes suspicious.
Suspicion is important.
Not paranoia. Not blind rebellion. Not childish defiance.
A disciplined suspicion.
The kind that says: before I accept this limit, I want to inspect the frame that produced it.
This is how revolutions begin. Not with noise, but with a better question.

THE AGE OF FLUENT NONSENSE

The old danger was ignorance.
The new danger is polished ignorance.
In the past, bad thinking often looked bad. It had rough edges. It sounded incomplete. It exposed itself through clumsy language, missing evidence, and visible confusion.
AI has changed this.
Now weak thinking can arrive in perfect paragraphs.
Falsehood can wear elegant formatting.
Assumptions can hide inside bullet points.
Speculation can sound like expertise.
A hallucination can walk into the room wearing a suit.
This is why verification is no longer optional.
It is not a bureaucratic exercise.
It is not a school habit.
It is not something only researchers, lawyers, doctors, engineers, or journalists need.
Verification is now a survival ritual for the thinking person.
Because every unverified answer that enters your mind becomes part of your internal operating system.
If you repeatedly consume unverified AI output, you are not just gathering information. You are training your brain to confuse fluency with truth.
That is how the brain gets sold for peanuts.
Not in one dramatic act.
Not through one foolish mistake.
But through daily surrender.
ONE ACCEPTED ANSWER.

ONE UNCHECKED CLAIM.
ONE COPIED SENTENCE.
ONE CONVENIENT ASSUMPTION.
ONE LAZY "SOUNDS RIGHT."
ONE MORE MOMENT WHERE THE MACHINE THINKS AND THE HUMAN MERELY NODS.
The Verification Ritual exists to interrupt that surrender.

TRUST IS NOT THE STARTING POINT

The mantra of this chapter is simple:
Trust is not the starting point. Verification is.
This does not mean we must become cynical about everything. Cynicism is lazy in its own way. The cynic rejects too quickly. The gullible person accepts too quickly. Both avoid the harder work of discernment.
Verification is different.
Verification says:
I will not worship the answer.
I will not reject the answer merely because it came from a machine.
I will examine it.
I will test it.
I will compare it.
I will challenge it.
Then I will decide what I think.
This is adult intelligence.
Not passive trust.
Not fashionable distrust.
But disciplined judgment.
AI should not be treated as an oracle. It should be treated as a powerful but unreliable collaborator: useful, fast, sometimes brilliant, sometimes wrong, often incomplete, always requiring supervision.
The user who trusts too quickly becomes a consumer.
The user who verifies becomes a thinker.
The user who verifies and reframes becomes a designer.

THE FIVE-STEP VERIFICATION RITUAL

Every important AI output should pass through a ritual.

Not every small task needs this level of rigor. You do not need a philosophical investigation to ask AI for a grocery list or a draft subject line. But when the output affects belief, money, strategy, health, education, reputation, creativity, law, design, or human decisions, the ritual must begin.

Treat the AI response like a crime scene.

Do not admire the carpet.

Look for footprints.

Step 1 — Ask: What is the claim?

The first task is to strip the answer naked.

AI often surrounds claims with confident language, comforting transitions, and polished explanation. Your job is to remove the perfume.

Ask:

What is this response actually claiming?

Which statements are facts?

Which are interpretations?

Which are guesses?

Which are recommendations?

Which are merely decorative?

A machine may say:

"This strategy will significantly improve user engagement."

That sounds impressive. But what is the claim?

Improve by how much?

For whom?

In what context?

Based on what evidence?

Compared to what previous condition?

Over what time period?

At what cost?

The first step of verification is precision.

A vague claim cannot be verified.

A precise claim can be tested.

So before you argue with an answer, sharpen it.

Step 2 — Check: What evidence supports it?

The second step is evidence.

Not vibes.

Not elegance.
Not confidence.
Not "studies show."
Not "experts agree."
Not "it is widely believed."
Evidence.
What study?
Which expert?
What data?
What source?
What year?
What sample size?
What context?
What limitation?
What competing interpretation?
AI is very good at sounding like evidence has already been handled. This is dangerous because the style of authority can replace the substance of authority.
The awakened mind refuses this substitution.
If an AI response gives a number, demand the source.
If it gives a historical claim, check the date.
If it gives a legal claim, verify the jurisdiction.
If it gives a scientific claim, inspect the research basis.
If it gives a business recommendation, ask what market conditions it assumes.
If it gives a design recommendation, ask what user reality it is based on.
Evidence is not decoration. Evidence is the spine.
Without evidence, the answer is only a well-dressed possibility.

Step 3 — Compare: What do independent sources say?
One source is a whisper.
Two sources are a conversation.
Multiple independent sources begin to create a map.
Comparison protects you from the prison of a single answer.
AI can produce a response that feels complete because it is linguistically complete. But reality does not become complete just because a paragraph ends neatly.
So compare.
Look beyond the machine.

Look at primary sources.
Look at opposing views.
Look at human expertise.
Look at lived experience.
Look at older sources and newer sources.
Look at what practitioners say when they are not writing marketing copy.

The goal is not to drown in research. The goal is to break the spell of the first answer.

The first answer is often the most dangerous because it gives emotional relief.

It says, "Here, you can stop now."

The Verification Ritual says, "Not yet."

Step 4 — Challenge: What could be false, missing, biased, or oversimplified?

This is where verification becomes design intelligence.

Checking evidence is not enough. You must also challenge the frame.

Ask:

What is missing?
Who is absent from this answer?
What does this response assume is normal?
What historical bias might be hidden inside it?
What has been simplified too aggressively?
What uncomfortable truth has been smoothed over?
What would a critic attack first?
What would a practitioner laugh at?
What would a user experience differently in the real world?

AI often produces the average of the known. But breakthrough thinking often begins at the edge of the known.

So you must ask the machine to expose its own blind spots.

A response may be factually decent but strategically dead.

It may be logical but unimaginative.
It may be efficient but unethical.
It may be popular but shallow.
It may be true in general but false in your context.

This is why challenge is necessary.

Verification is not only about catching lies. It is also about catching smallness.

Step 5 — Synthesize: What do I now think?

This is the step many people skip.

They ask AI.

They check a little.

Then they still let the AI's conclusion become their conclusion.

That is not thinking. That is outsourced opinion laundering.

Synthesis means the final judgment returns to the human mind.

After asking, checking, comparing, and challenging, you must say:

What do I now believe?

What is still uncertain?

What has changed in my view?

What should be rejected?

What should be kept?

What should be redesigned?

What is the new question?

This is where the human becomes sovereign again.

AI can provide material.

AI can provide options.

AI can provide friction.

AI can provide counterarguments.

But the final synthesis must belong to the awakened mind.

Otherwise, the machine has not assisted your thinking. It has replaced it.

FROM VERIFICATION TO REFRAMING

Verification protects the mind from falsehood.

But reframing protects the mind from smallness.

This is the next move.

Once you verify an answer, you must ask whether the entire question was too small.

For example, suppose you ask:

"How can I make employees more productive?"

AI may give you a list of productivity tools, meeting policies, automation workflows, and performance metrics.

Fine.

Now verify the claims.

But then reframe the question.

Maybe the real question is not:
"How do I make employees more productive?"
Maybe the real question is:
"Why is so much human energy being wasted inside this system?"
Or:
"What kind of work deserves human attention?"
Or:
"Which tasks should disappear instead of being optimized?"
Or:
"What would a humane productivity system look like?"
This is the difference between management thinking and design thinking.
Management often improves the machine.
Design questions whether the machine deserves to exist in its current form.
The same applies to AI.
A peanut question asks:
"How can AI help me produce more content?"
A paradigm question asks:
"What content is worth producing in a world already drowning in content?"
A peanut question asks:
"How do I get faster answers?"
A paradigm question asks:
"Which questions deserve slowness?"
A peanut question asks:
"How do I automate my thinking?"
A paradigm question asks:
"What parts of thinking must never be automated?"
This is where impossibility turns into a design brief.

TURNING "IMPOSSIBLE" INTO A DESIGN BRIEF

When someone says, "That is impossible," do not stop.
Translate the statement.
The word "impossible" usually hides a more useful sentence.
"It is impossible" may mean:
We do not have the budget.

We do not have the time.
We do not have the skill.
We do not have the language.
We do not have the courage.
We do not have the authority.
We do not have the evidence.
We do not have the imagination.
We do not have the right frame.
Each of these creates a different design brief.
If the barrier is budget, the brief becomes:
How might we achieve the core value with radically fewer resources?
If the barrier is time, the brief becomes:
What is the smallest meaningful version we can test quickly?
If the barrier is skill, the brief becomes:
What capability must be learned, borrowed, partnered, or simplified?
If the barrier is authority, the brief becomes:
Who must be persuaded, bypassed, or included?
If the barrier is evidence, the brief becomes:
What experiment would reduce uncertainty?
If the barrier is imagination, the brief becomes:
What assumptions are making this feel impossible?
This is the designer's move.
Do not argue with impossibility in its raw form. Translate it into constraints. Then redesign around the constraints.
A constraint is not an enemy.
A constraint is a shape.
The designer does not need infinite freedom. In fact, infinite freedom often produces weak thinking. Constraints force intelligence to become specific.
The question is not, "How do I remove every constraint?"
The better question is:
Which constraint is real, which is inherited, and which can become creative fuel?

THE FRICTION ARCHITECTURE

Before the Verification Ritual can work, the mind must be trained

to tolerate friction.
This is difficult because modern life has made friction feel like failure.
Waiting feels like failure.
Doubt feels like failure.
Revision feels like failure.
Not knowing feels like failure.
Changing your mind feels like failure.
But these are not failures.
They are signs that thinking is actually happening.
The lazy brain wants instant closure.
The awakened brain can remain open under pressure.
That is why we need a Friction Architecture: a set of habits that protects the mind from becoming too smooth, too obedient, too easily satisfied.

Habit 1 — The Fifteen-Minute Blank Void

Before touching a screen, sit with a blank page.
No notifications.
No feeds.
No AI.
No borrowed opinions.
No algorithmic weather entering your mind before your own climate has formed.
For fifteen minutes, write.
Write badly.
Write honestly.
Write fragments.
Write anxieties.
Write questions.
Write images.
Write contradictions.
Write what you do not yet understand.
This practice is simple, but it is not easy.
The brain will resist. It will ask for input. It will crave stimulation.
It will want the machine to begin on its behalf.
Do not let it.
This is how you reclaim authorship.
If the first voice you hear every morning is the algorithm, your

own voice becomes secondary.
The blank page is not empty. It is a mirror.

Habit 2 — The Rebel Sparring Match
Never use AI only to agree with you.
Agreement is cheap.
Resistance is valuable.
Take an idea you care about and ask AI to attack it.
Ask:
What is weak here?
What am I refusing to see?
What would my harshest critic say?
What assumptions must be true for this to work?
Where could this collapse?
Why might this be mediocre?
What would make this idea dangerous, boring, or irrelevant?
Do not do this because AI is always right.
Do it because friction strengthens structure.
A good idea should survive attack.
A great idea should improve because of attack.
This is not negativity. This is design pressure.
The goal is not to destroy your idea. The goal is to remove the parts that cannot carry weight.

Habit 3 — The Linguistic Detox
AI has a smell.
Not always, but often.
It appears in velvet phrases, inflated transitions, generic uplift, and sentences that sound intelligent but belong to no living person.
Words like:
"Delve."
"Testament."
"Moreover."
"It is important to note."
"In today's rapidly evolving landscape."
"Unlock the potential."
"A rich tapestry."
"Seamlessly integrates."
"Transformative journey."

These phrases are not evil. But when they appear without necessity, they reveal a deeper problem: the voice has gone soft.

Language is not only communication. It is evidence of thought. When your language becomes generic, your thinking often follows.

So detox.

Delete sentences that sound like they were grown in a corporate greenhouse.

Replace them with language that has bone, breath, and fingerprints.

Write like someone who has actually lived inside the problem.

Write with grit.

Write with risk.

Write with your own weather.

This matters because a person who cannot hear the difference between a human sentence and a machine-scented sentence will eventually lose the ability to hear himself.

PROMPTING AS INTELLECTUAL COMBAT

Most people prompt AI like they are ordering fast food.

"Give me this."

"Make it shorter."

"Make it professional."

"Write an article."

"Summarize this."

"Create a plan."

There is nothing wrong with utility. But if prompting never rises above extraction, the user remains mentally passive.

The stronger mode is provocation.

Use AI to provoke your thinking, not replace it.

Ask it to expose assumptions.

Ask it to separate facts from speculation.

Ask it to create falsification checklists.

Ask it to argue from opposing schools of thought.

Ask it to find where your frame is too small.

Ask it to produce uncomfortable questions.

Ask it to identify the hidden cost of your preferred solution.

Useful prompts include:

"List the assumptions behind your answer. Which ones are fragile?"

"Separate this response into facts, inferences, and speculation."
"What might be wrong, missing, biased, or oversimplified here?"
"Give me three opposing interpretations of the same situation."
"Create a falsification checklist. What would prove this answer wrong?"
"Act as a hostile reviewer. Attack this idea brutally but fairly."
"What question should I have asked instead?"
That last prompt is especially powerful.
What question should I have asked instead?
This turns prompting into reframing.
And reframing is where design begins.

THE BARNABY PROBLEM: ELEGANT COLLAPSE

Let us return to Barnaby.
Barnaby is launching an organic snack brand. He is excited, slightly anxious, and deeply allergic to difficult thinking. So he opens an AI tool and types:
"Give me a foolproof financial projection for an artisanal oat-bar company."
The machine responds beautifully.
There are charts.
There are growth curves.
There are margins.
There are market segments.
There are investor-friendly phrases.
There is a five-year projection so smooth it looks like destiny.
Barnaby feels relief.
This is the first warning sign.
Relief is not proof.
Relief is often what bad thinking uses as bait.
Barnaby takes the projection to investors. During the presentation, one investor asks:
"Your model assumes organic oat prices will fall by 40% next year. Where did that assumption come from?"
Barnaby freezes.
He did not ask what claims were being made.
He did not check the evidence.
He did not compare agricultural forecasts.
He did not challenge the assumptions.

He did not synthesize his own view.
He outsourced confidence.
The AI did not fail loudly. It failed elegantly.
That is the terrifying part.
A crude mistake might have alerted him. But a polished mistake seduced him.
This is how fluent nonsense becomes real-world damage.
Not because the machine is evil.
Because the human was asleep.
Barnaby did not lose because AI hallucinated.
He lost because he treated hallucination as strategy.

THE DESIGNER DOES NOT WORSHIP THE FIRST FRAME

The designer's real skill is not merely creativity.
It is frame discipline.
A frame determines what counts as a problem, what counts as evidence, what counts as success, and what kinds of solutions are allowed to appear.
Most people live inside inherited frames.
They inherit frames from school.
From industry.
From family.
From software.
From job titles.
From culture.
From dashboards.
From templates.
From the language everyone around them keeps using.
AI often reinforces these inherited frames because it is trained on what has already been said and done.
This does not make AI useless. It makes the human role more important.
The designer must ask:
What frame is this answer assuming?
What frame is this industry trapped inside?
What frame would make the impossible possible?
What frame would make the current solution unnecessary?

What frame would turn this from a problem into a redesign opportunity?

This is the difference between using AI as a clerk and using AI as a creative adversary.

The clerk fills the form.

The adversary exposes the form.

FROM PEANUTS TO PARADIGMS

The title of this book carries a warning: Don't sell your brain for peanuts.

Peanuts are the small bargains we make with convenience.

We sell our attention for noise.

We sell our judgment for speed.

We sell our imagination for templates.

We sell our language for generic polish.

We sell our curiosity for summaries.

We sell our responsibility for automation.

The transaction feels harmless because each peanut is small.

But over time, these small bargains accumulate into a life of cognitive dependency.

You wake up and ask the machine what to think.

You write and let the machine choose your voice.

You design and let the machine define the options.

You decide and let the machine hide the assumptions.

You create and let the machine reduce your imagination to averages.

That is the peanuts condition.

Paradigms are different.

A paradigm is not just an idea. It is a new way of seeing.

When the frame changes, the possible changes.

A paradigm mind does not merely ask for better answers. It builds better questions.

IT DOES NOT MERELY SEEK PRODUCTIVITY. IT SEEKS MEANING.

IT DOES NOT MERELY AUTOMATE THE OLD. IT IMAGINES THE NECESSARY.

IT DOES NOT MERELY ACCEPT "IMPOSSIBLE." IT INTERROGATES THE FRAME THAT PRODUCED THE WORD.

This is the journey of the book.

From mental laziness to cognitive sovereignty.

From hallucination to verification.

From offloading to disciplined augmentation.
From status quo repetition to frame-breaking imagination.
From user to thinker.
From thinker to designer.
From peanuts to paradigms.

THE VERIFICATION RITUAL AS CREATIVE FIRE

At first, verification may sound defensive.
Check the facts.
Inspect the claims.
Challenge the bias.
Avoid being fooled.
But verification is more than defense.
Verification is creative fire.
WHEN YOU VERIFY, YOU DISCOVER WEAK ASSUMPTIONS.
WHEN YOU DISCOVER WEAK ASSUMPTIONS, YOU FIND NEW OPENINGS.
WHEN YOU FIND NEW OPENINGS, YOU CAN REDESIGN THE BRIEF.
WHEN YOU REDESIGN THE BRIEF, IMPOSSIBILITY BEGINS TO CRACK.
This is why verification belongs in the creative heart of the book.
It is not the enemy of imagination.
It is the furnace that makes imagination strong.
An unverified imagination floats.
A verified imagination cuts.
An unverified dream inspires for a moment.
A verified dream can become architecture.
An unverified idea seduces.
A verified idea survives.
The great designer is not the person who dreams without limits.
The great designer is the person who knows which limits are real, which are false, and which can be transformed into form.

THE IMPOSSIBLE BRIEF METHOD

Whenever you face an impossible problem, use this method.

1. Name the wall

State the impossibility clearly.
Not vaguely. Clearly.
"We cannot build this product within the current budget."

"We cannot trust this AI output without verification."
"We cannot teach creativity through memorization alone."
"We cannot scale this process without losing quality."
"We cannot solve this problem using the current frame."

Naming the wall prevents emotional fog.

2. Identify the frame

Ask:

What assumptions make this impossible?
What rules are being treated as permanent?
What definitions are too narrow?
What success metric is distorting the solution?
What old model is controlling the conversation?

3. Separate real constraints from inherited constraints

Some constraints are real. Respect them.

Some constraints are inherited. Question them.

A real constraint says: "This material cannot carry that load."
An inherited constraint says: "We have always done it this way."
A real constraint says: "This law currently prohibits that action."
An inherited constraint says: "People like us do not try that."
A real constraint requires design intelligence.
An inherited constraint requires rebellion.

4. Convert the wall into a brief

Turn the impossibility into a design challenge.

Instead of:

"This is impossible."

Ask:

"How might we make the smallest testable version?"
"How might we solve the deeper need another way?"
"How might we remove the assumption creating the blockage?"
"How might we use AI without surrendering judgment?"
"How might we create a system where the human mind becomes stronger, not weaker?"

5. Prototype the new frame

Do not wait for perfect certainty.

Build a sketch.

Create a draft.

Run a test.

Make a model.

Write a counterargument.

Try the strange version.
Ask the forbidden question.
Paradigms do not arrive fully dressed.
They begin as rough prototypes.

WHY THIS CHAPTER MATTERS

This chapter matters because the entire book has been moving toward this point.
The early chapters warned against the flimsy mind, the lethargic user, and the outsourced self. They showed how easily human beings trade effort for comfort and call that trade intelligence.
Then the book examined AI as a status quo machine: powerful, useful, but often bound to the patterns of the past.
Now this chapter gives the response.
Do not reject AI.
Do not worship AI.
Do not lazily consume AI.
Do not let AI become the author of your assumptions.
Verify.
Challenge.
Reframe.
Design.
This is the ritual.
This is how the brain stays alive.

CLOSING REFLECTION: THE SACRED PAUSE BEFORE BELIEF

The most important moment in the age of AI may be the pause after the answer appears.
The machine has spoken.
The paragraph looks clean.
The formatting is elegant.
The anxiety drops.
The temptation arrives:
Accept it.
Use it.
Move on.
But the awakened mind pauses.

In that pause, sovereignty returns.
The pause says:
Not yet.
What is the claim?
What is the evidence?
What do other sources say?
What is missing?
What do I now think?
What frame is trapping this answer?
What would make the impossible possible?
That pause is not hesitation.
It is resistance.
It is the difference between being fed and being awake.
The future will not belong to the people who ask AI the most questions. It will belong to those who ask the deepest questions, verify the answers, and redesign the frames.
Impossible is not always a wall.
Sometimes it is a door with bad instructions.
The designer's task is to rewrite the instructions.
The thinker's task is to verify the door exists.
The creator's task is to walk through and build the next room.
That is the journey from peanuts to paradigms.
Not merely using the brain.
Not merely protecting the brain.
But awakening it enough to redesign the possible.

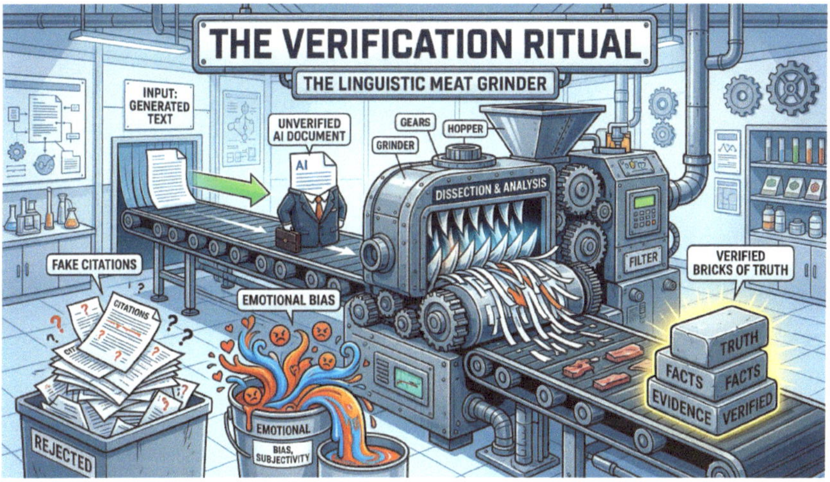
THE VERIFICATION RITUAL
THE LINGUISTIC MEAT GRINDER
INPUT: GENERATED TEXT
UNVERIFIED AI DOCUMENT
AI
GEARS
HOPPER
GRINDER
DISSECTION & ANALYSIS
FILTER
VERIFIED BRICKS OF TRUTH
FAKE CITATIONS
CITATIONS
EMOTIONAL BIAS
EMOTIONAL
BIAS, SUBJECTIVITY
REJECTED
TRUTH
FACTS
FACTS
EVIDENCE
VERIFIED

Chapter 10 From Prompting to Provocation

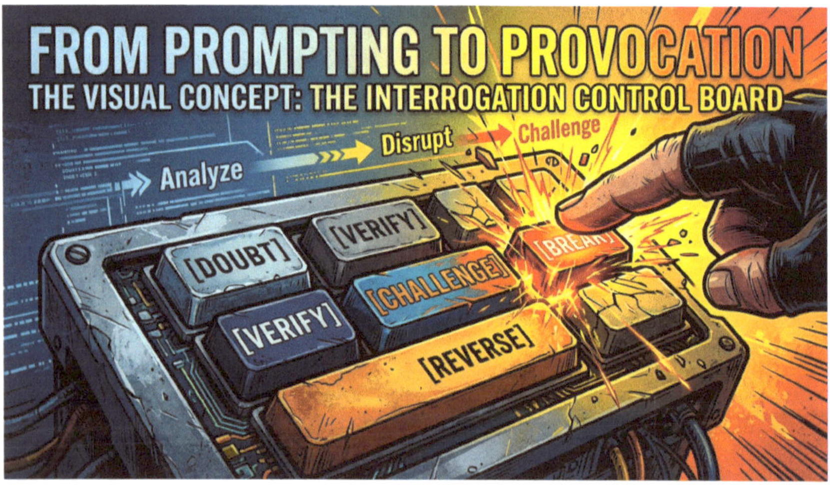

Stop Asking for Answers. Start Asking Better Questions.

Most people think a prompt is something you type into a box.
That is the first mistake.
A prompt is not merely a request.
A prompt is not a command.
A prompt is not a polite digital instruction whispered into the ear of a machine.
A prompt is a visible form of thinking.
It reveals the size of your question, the quality of your doubt, the sharpness of your intention, and the weakness or strength of your mental frame.
A lazy prompt exposes a lazy mind.
A vague prompt exposes a vague problem.
A shallow prompt exposes shallow hunger.
A copied prompt exposes borrowed thinking.
But a strong prompt does something different.
IT PROVOKES.
IT FORCES CONTRADICTION TO SURFACE.
IT DRAGS ASSUMPTIONS INTO THE LIGHT.
IT DEMANDS ALTERNATIVES.
IT CHALLENGES THE OBVIOUS ANSWER.
IT REFUSES THE COMFORT OF FIRST CLOSURE.
IT DOES NOT ASK AI TO FINISH THOUGHT. IT FORCES THOUGHT TO BEGIN.
This is the core shift of this chapter:
Weak prompts ask AI to finish thought. Strong prompts force thought to begin.
That one sentence separates the lethargic user from the awakened designer.

THE PROMPT BOX IS A MIRROR

The blank prompt box is not empty.
It is a mirror.

When you sit before it, you are not only interacting with software. You are encountering the current condition of your own mind.
If you type:
"Write me something about innovation."
The problem is not only the prompt.
The problem is the thought behind the prompt.
What kind of innovation?
For whom?
Against what constraint?
Inside what broken system?
With what ethical tension?
What is the hidden failure you are trying to repair?
What assumption are you too lazy to question?
The machine cannot rescue a mind that refuses to define its own hunger.
AI may still give you something. It will always give you something. That is part of the danger. It may produce smooth paragraphs, clean headings, and a tone of synthetic confidence. But beneath that surface, the answer will often be a statistical fog produced from an unclear human request.
A weak prompt says:
"Complete this for me."
A strong prompt says:
"Disturb me. Challenge me. Show me what I cannot yet see."
That is where prompting becomes provocation.

The Age of Answer Addiction
Modern people are addicted to answers.
Not truth.
Not wisdom.
Not transformation.
Answers.
Answers reduce discomfort.
Answers create closure.
Answers make us feel temporarily intelligent.
Answers allow us to move on without changing.
This addiction existed before AI, but AI has industrialized it.
Now the answer arrives instantly.

Formatted.
Confident.
Friendly.
Ready to paste.
Ready to submit.
Ready to publish.
Ready to believe.
The danger is not that AI gives answers.
The danger is that people may forget that an answer is not the same as understanding.
An answer can be borrowed.
Understanding must be built.
An answer can be generated.
Understanding must be wrestled into existence.
An answer can be copied in seconds.
Understanding demands friction, doubt, revision, contradiction, and synthesis.
The lethargic user wants AI to remove the pain of thinking.
The designer uses AI to increase the quality of thinking.
That is the rebellion.

STOP ASKING FOR COMFORT

Many prompts are emotional rescue missions in disguise.
The user is anxious, confused, behind schedule, insecure, or mentally tired. So he asks AI to remove the tension.
"Give me the best answer."
"Write this quickly."
"Make this sound professional."
"Summarize everything."
"Tell me what to do."
The machine responds. The user feels relief.
But relief is not the same as insight.
Sometimes relief is the sugar coating on mental surrender.
The best thinking does not always reduce tension immediately.
Sometimes it increases tension first. It reveals that the problem is bigger than expected, the assumptions are weaker than believed, the evidence is thinner than hoped, and the first solution is too small.

That is not failure.
That is thinking waking up.
A great prompt does not always make life easier. It makes the mind sharper.
It asks AI to become a sparring partner, not a pillow.

THE WEAK PATH AND THE MAVERICK PATH

There are two ways to use AI.
The first is the weak path.

Anxiety → Vague Prompt → Instant Answer → Temporary Relief → Cognitive Decay

Anxiety → Vague Prompt → Instant Answer → Temporary Relief → Cognitive Decay

This path feels productive. It is not.
It produces motion without depth. It creates output without ownership. It gives the user the illusion of intelligence while quietly weakening the muscles that create intelligence.
The second is the maverick path.

Curiosity → Provocative Prompt → Productive Tension → Reframing → Original Insight

Curiosity → Provocative Prompt → Productive Tension → Reframing → Original Insight

This path is slower at first. It is less comfortable. It may even feel irritating because it refuses to flatter you.
But it changes the user.
The weak path asks AI to do the work.
The maverick path asks AI to make the human more dangerous.
Dangerous to shallow thinking.
Dangerous to false assumptions.
Dangerous to inherited frames.
Dangerous to the status quo machine.
Dangerous to the part of the self that wants comfort more than truth.
This is the point of provocation.

PROMPTING IS NOT A TECHNICAL SKILL FIRST

There are many guides that teach people how to write better prompts.

Give context.
Define the role.
Specify the format.
Add examples.
State the audience.
Ask for step-by-step reasoning.
Iterate.
These are useful techniques.
But they are not the heart of prompting.
The heart of prompting is not technical. It is intellectual.
A good prompt begins before the typing begins.
It begins with the human asking:
What am I actually trying to understand?
What am I assuming?
Where could I be wrong?
What kind of answer would be too easy?
What tension am I avoiding?
What would change if I asked the question differently?
Without this inner work, prompt engineering becomes another form of decoration.
You may learn the syntax of prompting and still remain a weak thinker.
You may master templates and still ask timid questions.
You may produce beautiful AI output and still never confront the real problem.
The future will not belong to the people with the longest prompt templates. It will belong to those with the strongest questions.

A PROMPT IS A DESIGN BRIEF FOR THOUGHT

A design brief shapes what can be created.
If the brief is small, the result will be small.
If the brief is confused, the result will be confused.
If the brief hides the real problem, the solution will solve the wrong thing beautifully.
A prompt works the same way.
A prompt is a design brief for thought.
It tells the machine what world to enter, what tension to examine,

what constraints to respect, what assumptions to inspect, and what kind of transformation is being requested.

A bad prompt says:

"Give me ideas."

A better prompt says:

"Give me ten ideas."

A stronger prompt says:

"Give me ten ideas that challenge the dominant assumption in this industry."

A provocative prompt says:

"Give me ten ideas that would seem unreasonable to an established expert but might become necessary if the current system collapses."

Now the machine is not merely filling space. It is being pushed into a frame-breaking role.

The user is no longer asking for content.

The user is designing a collision.

FROM ANSWER-SEEKING TO TENSION-SEEKING

The average user asks:

"What is the answer?"

The designer asks:

"What tension is hidden inside the question?"

This changes everything.

For example, a weak prompt asks:

"Write an article about AI and creativity."

A stronger prompt asks:

"Write an article about the tension between AI-generated creativity and human originality."

A provocative prompt asks:

"Argue that AI does not kill creativity directly, but seduces humans into abandoning the discomfort from which originality is born. Then challenge that argument from the opposite side."

Now the prompt has teeth.

It does not merely request material. It creates intellectual pressure.

Tension is where insight lives.

Between speed and depth.

Between automation and ownership.
Between logic and imagination.
Between evidence and intuition.
Between convenience and sovereignty.
Between what is probable and what is necessary.
A weak mind avoids these tensions.
A strong mind enters them deliberately.

THE PROVOCATIVE PROMPT FORMULA

A strong provocative prompt often contains five parts:

Context + Tension + Challenge + Constraint + Desired Transformation

Context + Tension + Challenge + Constraint + Desired Transformation

1. Context

What is the situation?

Do not make AI guess the world you are working inside. Give it the terrain.

Example:

"I am developing a chapter for a book about how AI can weaken critical thinking when people use it lazily."

2. Tension

What conflict must be explored?

Example:

"The tension is between using AI as a tool for augmentation and using AI as an excuse to stop thinking."

3. Challenge

What should the AI attack, expose, reverse, or test?

Example:

"Challenge the assumption that faster output is always better."

4. Constraint

What boundaries should guide the response?

Example:

"Use a gritty, philosophical style with strong examples, but avoid empty corporate language."

5. Desired Transformation

What should change in the reader's mind?

Example:

"The reader should stop seeing prompting as typing and start see-

ing it as a discipline of thought."
This is prompting as design.
You are not merely asking for a response.
You are shaping the conditions under which thinking can become more alive.

THE EIGHT KEYS OF PROVOCATION

Imagine a keyboard built not for typing, but for awakening the mind.
Each key is not a letter. It is an intellectual action.

[DOUBT] [TEST] [REVERSE] [COMPARE]

[IMAGINE][VERIFY][BREAK] [REBUILD]

[DOUBT] [TEST] [REVERSE] [COMPARE]

[IMAGINE][VERIFY][BREAK] [REBUILD]

This is the designer's control board.

DOUBT

Doubt is not weakness. Doubt is the beginning of precision.
Prompt:
"What assumptions in this idea should I doubt first?"

TEST

Testing prevents imagination from becoming fantasy.
Prompt:
"Create a practical test that would reveal whether this idea has real value."

REVERSE

Reversal breaks the tyranny of the obvious.
Prompt:
"Reverse the usual assumption behind this problem. What becomes possible?"

COMPARE

Comparison prevents single-answer blindness.
Prompt:
"Compare three opposing ways to interpret this situation."

IMAGINE

Imagination expands the frame.

Prompt:
"Imagine a solution that would look strange today but necessary ten years from now."

VERIFY

Verification protects the mind from fluent nonsense.
Prompt:
"Separate verified facts, inferences, assumptions, and speculation."

BREAK

Breaking exposes hidden structure.
Prompt:
"Break this idea. Where will it fail under pressure?"

REBUILD

Rebuilding turns critique into creation.
Prompt:
"Now rebuild the idea so it survives the strongest criticism."
This is not a keyboard for passive users.
It is an instrument for cognitive combat.

THE LAZY PROMPT AND THE LIVING PROMPT

A lazy prompt asks for completion.
A living prompt creates movement.

Lazy prompt:

"Give me a business idea."

Living prompt:

"Give me a business idea that challenges a hidden inefficiency in how people currently waste attention, then explain why the idea might fail."

Lazy prompt:

"Write a speech about leadership."

Living prompt:

"Write a speech arguing that modern leadership fails when it confuses confidence with clarity. Include the strongest counterargument."

Lazy prompt:

"Summarize this article."

Living prompt:

"Summarize this article, identify its three weakest assumptions, and suggest what the author is refusing to confront."

Lazy prompt:

"Make this better."

Living prompt:

"Make this sharper by removing generic language, increasing conceptual tension, and preserving my original voice."

Lazy prompt:

"What should I do?"

Living prompt:

"Give me three possible decisions, the assumptions behind each, the risks I am likely underestimating, and the question I should ask before choosing."

A lazy prompt ends the conversation.

A living prompt opens the wound properly so healing can begin.

THE PROMPT AS X-RAY

A strong prompt should function like an X-ray.

It should reveal hidden bones.

Every idea has a skeleton: assumptions, incentives, contradictions, fears, blind spots, dependencies, and consequences.

Most people only look at the skin of an idea.

Does it sound good?

Does it look professional?

Does it feel exciting?

Does it match what others are saying?

The designer looks deeper.

What is holding this idea upright?

Where is the fracture?

What load can it carry?

What pressure will break it?

What is hidden beneath the attractive surface?

AI can help expose this skeleton if you force it to.

Ask:

"What must be true for this idea to work?"

"What would make this idea collapse?"

"What hidden dependency is not obvious?"

"What contradiction is being covered by good language?"

"What does this idea assume about human behaviour?"

These questions turn AI from a content generator into a diagnostic instrument.

But the instrument only works when the human knows what to inspect.

THE STRONGEST OPPOSING CASE

One of the most powerful uses of AI is to force opposition.

Most people use AI to defend what they already believe. That is comforting, but dangerous.

The awakened user asks AI to attack.

Not weakly. Not politely. Not with artificial balance.

Strongly.

Prompt:

"Construct the strongest possible argument against my position. Do not use straw-man objections. Make the opposing case intelligent, fair, and difficult to dismiss."

This prompt is a cure for intellectual vanity.

If your idea cannot survive its strongest opponent, it is not ready.

This does not mean you must abandon your idea every time it is challenged. It means you must improve it until it can stand under pressure.

Weak thinkers avoid opposition because it threatens identity.

Strong thinkers seek opposition because it strengthens structure.

The goal is not to win an argument.

The goal is to stop being fragile.

PROMPTING AGAINST THE STATUS QUO MACHINE

AI is often trained on what has already been said.

That means it can easily become a status quo machine.

Ask it a conventional question and it will often give you a conventional answer. Ask it for "best practices" and it may give you the average habits of yesterday. Ask it for "industry standards" and it may reproduce the mental furniture of the existing system.

This is not always bad. Sometimes standards exist for good reason.

But if you want paradigm shifts, you cannot stop at best practices.

Best practices often mean:

"What everyone respectable already accepts."
The designer must ask:
What comes after best practice?
What is the hidden cost of this best practice?
What practice only exists because the system is afraid to change?
What would a beginner question that an expert no longer sees?
What would an outsider notice immediately?
Provocative prompt:
"Give me the standard answer to this problem. Then attack the standard answer. Then propose a frame-breaking alternative that would make experts uncomfortable but users relieved."
Now AI is not merely repeating the past.
It is being forced to expose the edges of the past.

PROMPTING FOR THE IMPOSSIBLE

After the Verification Ritual, the next discipline is the Impossible Prompt.
This is where the designer asks AI not only for what is likely, but for what might become necessary.
Weak prompt:
"How can we improve education?"
Impossible prompt:
"Assume the current education system is optimized for compliance more than curiosity. Redesign the learning experience around wonder, self-questioning, and verification. What would need to change first?"
Weak prompt:
"How can businesses use AI?"
Impossible prompt:
"Design a business workflow where AI increases human judgment instead of replacing it. What rituals, safeguards, and provocations must be built into the system?"
Weak prompt:
"How can I be more productive?"
Impossible prompt:
"Challenge the idea that my problem is productivity. What if the real problem is attention fragmentation, weak priorities, or fear of

deep work?"
This is how impossible becomes a design brief.
The prompt does not accept the surface problem.
It searches for the deeper one.

THE QUESTION BENEATH THE QUESTION

Every prompt has a question beneath the question.
The visible question may be:
"How do I write faster?"
The hidden question may be:
"Why am I afraid to write badly first?"
The visible question may be:
"How do I make this presentation impressive?"
The hidden question may be:
"Do I actually have a strong idea, or only decorative slides?"
The visible question may be:
"How do I use AI better?"
The hidden question may be:
"How do I stop using AI as a substitute for courage?"
A provocative prompt hunts the hidden question.
Ask AI:
"What is the deeper question underneath my question?"
This is one of the most important prompts in the age of artificial intelligence.
Because people often ask AI to solve the symptom, not the disease.
They ask for a better answer when they need a better frame.
They ask for polish when they need substance.
They ask for efficiency when they need direction.
They ask for confidence when they need verification.
They ask for completion when they need confrontation.
The deeper question is where the real work begins.

THE HUMAN MUST REMAIN THE FINAL EDITOR OF MEANING

AI can generate alternatives, but it cannot decide what should matter to you.
It can list priorities, but it does not carry your consequences.

It can simulate arguments, but it does not live with your choices.
It can produce language, but it does not own your voice.
That is why the human must remain the final editor of meaning.
This is not about ego.
It is about responsibility.
The person who publishes the thought is responsible for the thought.
The person who acts on the recommendation is responsible for the action.
The person who uses the tool is responsible for the consequences of using the tool lazily.
A weak user hides behind the machine.
"AI said this."
"AI gave me this."
"AI made it sound right."
That is not defense. That is confession.
The awakened user says:
"I used the machine. Then I verified. Then I challenged. Then I synthesized. Then I chose."
That is ownership.

THE PROMPTING LADDER

There is a ladder of prompting maturity.

Level 1 — The Request

"Write this."
This is the lowest level. Useful for minor tasks, dangerous for major thought.

Level 2 — The Instruction

"Write this for this audience in this tone."
Better. The user has added context and direction.

Level 3 — The Collaboration

"Help me explore options and compare approaches."
Now AI becomes a thinking partner.

Level 4 — The Critique

"Attack this idea and expose its weaknesses."
Now AI becomes a sparring partner.

Level 5 — The Reframe

"What question should I be asking instead?"
Now AI becomes a frame-breaker.

Level 6 — The Provocation

"Force me to confront the assumption that makes my current thinking too small."
Now the prompt is no longer a request.
It is an intellectual event.
This is where prompting becomes provocation.

THE DESIGNER'S PROMPTING RITUAL

A designer should not begin with:
"What can AI give me?"
The designer begins with:
"What must I think through before AI touches this?"
Then the ritual begins.

1. State the problem in your own words

Before asking AI, write your raw understanding.
This protects you from becoming mentally overwritten by the machine's first answer.

2. Identify your current assumption

Name what you already believe.
If you do not name your assumption, you cannot challenge it.

3. Ask for contradiction

Force the model to oppose you.

4. Ask for alternatives

Demand more than one path.

5. Ask for verification

Separate facts from speculation.

6. Ask for reframing

Find the better question.

7. Synthesize in your own voice

Do not end with AI output.
End with human judgment.
This ritual transforms prompting from convenience into discipline.

THE DEATH OF THE COPY-PASTE MIND

The copy-paste mind is one of the great tragedies of the AI age.
It does not look stupid. In fact, it may look very productive.
It produces documents.
It generates emails.
It posts articles.
It fills slides.
It creates strategies.
It sounds busy.
But underneath the activity, something has died.
The person is no longer wrestling with thought.
They are moving text from one surface to another.
Copy from machine.
Paste into document.
Copy from document.
Paste into world.
Repeat.
This is not creation.
This is clerical ventriloquism.
The voice appears, but the soul is absent.
The antidote is provocation.
A provocative prompt prevents copy-paste decay because it creates friction before output. It forces the user to examine, reject, reshape, and own the result.
The goal is not to produce more.
The goal is to become more awake while producing.

THE NEW PROMPTING VOCABULARY

Replace passive prompts with active verbs.
Do not merely ask AI to write.
Ask it to:

INTERROGATE.
DOUBT.
REVERSE.
STRESS-TEST.
COMPARE.
EXPOSE.

CONTRADICT.
SIMPLIFY.
COMPLICATE.
REFRAME.
VERIFY.
PROTOTYPE.
DIAGNOSE.
DISRUPT.
REBUILD.
HUMANIZE.
SHARPEN.
FALSIFY.
TRANSLATE.
MAP.
SIMULATE.
PRESSURE-TEST.

Each verb changes the relationship.

"Write" makes the machine a clerk.
"Challenge" makes it a critic.
"Verify" makes it an auditor.
"Reframe" makes it a designer.
"Falsify" makes it a scientist.
"Humanize" makes it a voice coach.
"Map" makes it a systems thinker.
"Break" makes it an engineer of failure.
"Rebuild" makes it a collaborator.

The verb you choose determines the intelligence you activate.

EXAMPLES OF PROVOCATIVE PROMPTS

For writing

"Read this paragraph and remove every generic phrase. Replace vague polish with sharper human language."

"What is the central tension in this chapter, and where am I avoiding it?"

"Find the sentence that sounds most like AI and rewrite it with grit."

For design

"What user pain is being normalized in this design instead of solved?"
"Which assumption in this interface would fail for a tired, distracted, or first-time user?"
"Redesign this experience as if dignity, not efficiency, were the primary metric."
For business
"What hidden cost does this business model push onto users, workers, or society?"
"What would make this strategy collapse in the real world?"
"Give me the strongest competitor response to this plan."
For learning
"Do not explain this immediately. First, ask me five questions that reveal what I misunderstand."
"Create a practice path that forces me to struggle productively instead of memorizing passively."
"What false confidence might a beginner develop after reading a shallow explanation of this topic?"
For creativity
"Give me three strange directions that break the obvious pattern."
"What cliché am I accidentally repeating?"
"Make this idea more original without making it obscure."
For verification
"Separate this answer into facts, assumptions, inferences, and speculation."
"What sources would I need to verify this properly?"
"Create a checklist to prove this answer wrong."
These are not magic spells.
They are thinking tools.
The magic is not in the wording. The magic is in the discipline behind the wording.

WHEN NOT TO PROMPT

The awakened user also knows when not to prompt.
This is important.
Not every mental struggle should be immediately handed to AI.
Sometimes the struggle is the point.

If you prompt too early, you may interrupt the formation of your own thought. You may let the machine's structure colonize your raw intuition before it has a chance to become original.

Before prompting, ask:

Have I tried to think this through myself?

Have I written my rough version?

Have I named my confusion?

Have I sat with the discomfort long enough?

Am I using AI to extend thought or avoid thought?

There is a sacred moment before input.

Protect it.

The mind needs some silence before it can ask a worthy question.

THE FINAL SHIFT: FROM USER TO PROVOCATEUR

The word "user" is too passive for the future.

A user operates a tool.

A provocateur changes the conversation.

A user asks for output.

A provocateur creates tension.

A user wants completion.

A provocateur wants transformation.

A user accepts the frame.

A provocateur interrogates the frame.

A user says, "Give me the answer."

A provocateur says, "Show me the contradiction that makes the answer too easy."

This is the identity shift at the heart of Chapter 10.

You are not here to become a better typist into a machine.

You are here to become a better thinker with a machine nearby.

That difference is everything.

CONCLUSION TO PART III — THE DESIGNER'S IMPOSSIBLE LENS

The Shift from Copier to Creator

Part III has moved through three awakenings.

Chapter 8 awakened the whole brain. It rejected the false war between logic and imagination. It argued that the future requires analysis plus wonder, evidence plus intuition, critique plus creative synthesis.

Chapter 9 gave us the Verification Ritual. It taught us not to trust fluency, not to worship polished answers, and not to confuse confidence with truth. It turned impossibility into a design brief and showed that many walls are only badly framed questions.

Chapter 10 completes the arc by transforming prompting itself.

A prompt is no longer a request.

It is a weapon of attention.
It is a mirror of thought.
It is a design brief for intelligence.
It is a lever for reframing reality.
It is a ritual for refusing mental sleep.

This is the journey from peanuts to paradigms.

The peanut mind wants shortcuts.
The paradigm mind wants transformation.
The peanut mind asks for answers.
The paradigm mind asks better questions.
The peanut mind accepts AI's first fluent offering.
The paradigm mind verifies, challenges, reframes, and rebuilds.
The peanut mind sells its judgment cheaply.
The paradigm mind protects its sovereignty.

The world does not need more copy-paste clerks wearing the costume of productivity. It does not need more polished summaries, more generic strategies, more synthetic confidence, more people who can generate but cannot judge.

The world needs designers of thought.

People who can doubt without becoming cynical.

Imagine without becoming foolish.

Verify without becoming sterile.

Prompt without becoming passive.

Use AI without becoming used by it.

That is the impossible lens.

It does not reject the machine.

It refuses to kneel before the machine.

It says:

I will use the tool, but I will not sell the mind.
I will seek assistance, but not surrender.
I will ask questions, but not shallow ones.
I will verify what looks fluent.

I will challenge what looks obvious.
I will reframe what looks impossible.
I will provoke the future instead of merely consuming the past.
The contract is broken.
The brain is no longer for sale.
Not for convenience.
Not for speed.
Not for pretty language.
Not for the peanuts of instant closure.
The mind returns to the workshop.
The designer sits before the blank page, the prompt box, the impossible problem, and the unfinished world.
And this time, he does not ask:
"Can you give me the answer?"
He asks:
"What question will break the frame?"

Part IV

Reclaiming Judgment

Flow Insight for Chapters 11–13

Part IV must feel like the recovery movement of the book. The earlier parts have exposed the danger: the flimsy mind, the lethargic user, hallucination, cognitive offloading, the status quo machine, and the peanut bargain. By now, the reader has seen how easily the human brain can be sold cheaply when comfort, speed, and convenience become substitutes for thought.

Now Part IV must not merely complain. It must restore authority.

Chapter 11 The Stewardship Mind

AI as Apprentice, Mirror, Critic, and Catalyst

We need to stop treating Artificial Intelligence as if it were an all-knowing digital deity that descended from the heavens to dictate human destiny.

It is not an oracle.

It is not your master.

It is not your moral compass.

It is not your imagination.

And it is certainly not a replacement for your soul.

AI is a tool. A remarkable tool, yes. A fast tool. A tireless tool. A tool that can draft, sort, summarize, mimic, compare, translate, visualize, and remix at a speed that makes the ordinary human workflow look prehistoric.

But speed is not wisdom.

Fluency is not truth.

Prediction is not judgment.

And convenience is not liberation.

The tragedy of our age is not that machines are becoming powerful. The tragedy is that many humans are becoming too willing to become weak in the presence of that power. They see a polished answer and mistake it for authority. They see confident language and mistake it for understanding. They see rapid output and mistake it for intelligence.

This is how the peanut bargain begins.

Not with a dramatic surrender. Not with a dystopian robot army. Not with a grand announcement that humanity has abdicated thought.

It begins quietly.

A student asks AI to think for him.

A designer accepts the first generated concept.

A manager forwards an AI summary without reading the source.

A writer lets the machine decide the emotional direction of the work.

A professional asks, "What should I believe?" instead of "What should I examine?"

Bit by bit, judgment is outsourced. Bit by bit, the brain is rented out for peanuts.

The Stewardship Mind is the refusal of that bargain.

It is the disciplined decision to use AI without being used by it.

It is the recognition that a powerful tool must be governed by a powerful human mind. Not a lazy mind. Not a frightened mind. Not a mind addicted to convenience. A stewarding mind.

The steward does not worship the tool.

The steward does not fear the tool.

The steward does not abandon responsibility to the tool.

The steward commands, tests, redirects, verifies, and finally decides.

That is the hierarchy this chapter restores.

AI may generate.

Humans must judge.

THE CROWN WAS NEVER THE MACHINE'S

The biggest mistake people make with AI is psychological before it is technical.

They place the machine above themselves.

They open the chat window like a confession booth. They ask the model to absolve them from the burden of thinking. They type their uncertainty into the box and wait for the glowing priest of probability to return an answer wrapped in calm, formatted paragraphs.

This is not assistance. This is intellectual submission.

The machine did not steal the crown.

The human removed it and handed it over.

That is why stewardship matters. Stewardship restores the proper chain of command.

The machine may know patterns, but the human knows stakes.

The machine may produce options, but the human must choose direction.

The machine may generate language, but the human must carry meaning.

The machine may simulate critique, but the human must endure consequence.

A steward understands that AI is not "above" the human mind.

It is beneath human responsibility. It sits on the workbench. It is a chisel, a compass, a microscope, a sketch assistant, a debating partner, a drafting engine.

It is not the hand.

It is not the eye.

It is not the conscience.

It is not the whole brain.

This distinction is not decorative. It is survival.

Because the moment you confuse tool-power with tool-authority, you begin to shrink. You begin to ask smaller questions. You begin to accept polished mediocrity. You begin to feel grateful for outputs that should have been raw material. You begin to call convenience "productivity," even when it is quietly draining your capacity to judge.

The Stewardship Mind says no.

It says:

I will use the machine's speed without surrendering my slowness.

I will use its range without surrendering my depth.

I will use its memory without surrendering my meaning.

I will use its fluency without surrendering my truth.

AI IS NOT ONE THING: ASSIGN IT ROLES

A major reason people misuse AI is that they treat it as a single vague intelligence.

They ask it anything and accept whatever comes back.

That is bad stewardship.

A good steward does not merely "use AI." A good steward assigns AI a role.

Role discipline is crucial because every role creates a boundary. Without boundaries, AI becomes a fog machine. It fills every gap with language. It answers before the question is mature. It smooths over contradictions before the human has wrestled with them. It provides closure where there should have been inquiry.

So we must stop treating AI as a universal answer box.

Instead, we assign it four controlled roles:

Role	Function	Boundary
Apprentice	Drafts, organizes, summarizes, formats	It does labor, not leadership
Mirror	Reflects assumptions, patterns, biases, defaults	It shows the frame, not the truth
Critic	Challenges logic, exposes weakness, tests claims	It attacks the argument, not the person
Catalyst	Generates alternatives, analogies, provocations	It sparks synthesis, but does not conclude

These roles are not gimmicks. They are governance structures.
They keep the machine in its place.
They keep the human awake.

---

1. AI AS APPRENTICE

The High-Speed Worker, Not the Master Designer
The first and safest role of AI is the Apprentice.
The Apprentice is fast, obedient, useful, and limited.
It can clean the table. It can sort the notes. It can draft the first version. It can summarize a long article. It can compare two frameworks. It can turn messy ideas into structure. It can convert raw fragments into a workable outline.
But the Apprentice does not own the vision.
This is where many people fail. They confuse first draft with final judgment. They treat AI's speed as evidence of superiority. They look at a neat structure and think the work is done.
No.
The Apprentice has only swept the floor.
The house has not been designed.
A strong human uses AI to reduce mechanical drag, not intellectual responsibility. There is nothing noble about wasting three hours formatting a table when AI can do it in thirty seconds.
There is nothing sacred about manually summarizing a document if the machine can produce a useful first-pass digest.
But there is something deeply dangerous about allowing the Ap-

prentice to decide what matters.

The Apprentice can prepare the clay.

The human must sculpt.

The Apprentice can assemble the bricks.

The human must design the cathedral.

The Apprentice can produce ten versions of a paragraph.

The human must know which sentence has blood in it.

This distinction saves the mind from both arrogance and laziness.

It allows us to accept help without surrendering authorship.

A good prompt to the Apprentice sounds like this:

"Organize these messy notes into clear themes. Do not add new claims. Preserve my voice. Mark gaps where more thinking is needed."

Notice the command. The human sets the boundary.

A weak user says:

"Write this for me."

A steward says:

"Prepare the material so I can think better."

That is the difference between offloading labor and offloading judgment.

THE APPRENTICE MUST BE SUPERVISED

An apprentice, by definition, is not independent.

The apprentice may be talented, but talent without supervision becomes chaos. AI can invent citations. It can misread the emotional tone of a subject. It can flatten nuance. It can produce clichés with the confidence of a statesman. It can sound correct while being quietly wrong.

This is why every AI-generated draft must pass through human inspection.

Ask:

- What has the machine assumed?
- What has it omitted?
- What has it made too smooth?
- Where is the emotional truth missing?
- Where does the writing sound impressive but empty?
- Where has it replaced my lived judgment with generic

phrasing?

The Apprentice is useful when supervised.

The Apprentice is dangerous when obeyed.

A steward never says, "AI wrote it, so it must be fine."

A steward says, "AI prepared it. Now I must examine it."

That shift alone protects the brain from peanut thinking.

---

2. AI AS MIRROR

The Reflection of the Status Quo

The second role of AI is the Mirror.

This is one of the most powerful and misunderstood uses of the technology.

AI is trained on human-produced patterns. It absorbs the dominant language of industries, cultures, professions, and ideologies. It knows what people usually say. It knows the shape of common answers. It can reproduce the consensus with frightening fluency.

That makes it dangerous when treated as truth.

But it makes it extremely useful when treated as a mirror.

The mirror does not reveal destiny.

It reveals conditioning.

When AI gives you a conventional answer, do not immediately accept it. Study it. Ask: What worldview is hiding inside this answer? What assumptions has it inherited? What does it consider normal? What does it refuse to imagine?

The Mirror helps you see the cage.

And once you see the cage, you can design the escape.

For example, a designer working on education might ask:

"What are the most common assumptions people make about the future of classrooms?"

AI may respond with predictable ideas: personalized learning, digital tutors, efficiency, gamification, remote access, performance tracking.

Useful.

Not because these ideas are original.

Useful because they show the default mental furniture of the field.

Now the designer can ask the real question:

"Which of these assumptions are lazy? Which ones preserve the existing system? Which ones pretend to innovate while keeping the same broken frame?"

That is stewardship.

The Mirror is not there to flatter your idea. It is there to reveal the wallpaper of conventional thought.

A weak user asks AI:

"What is the best solution?"

A steward asks:

"What would the average industry answer be, and how do I move beyond it?"

This is where the journey from peanuts to paradigms begins.

Peanut thinking accepts the mirror.

Paradigm thinking interrogates the reflection.

THE MIRROR ALSO SHOWS YOU YOURSELF

The Mirror does not only reflect society. It can reflect the user.

When you feed AI your outline, argument, design concept, or strategy, it can help expose your repeated patterns.

It may show that your language is too abstract.

It may show that your argument depends on a hidden assumption.

It may show that you keep circling the same emotional point without developing it.

It may show that your proposal sounds bold but lacks operational detail.

It may show that your "new" idea is actually wearing old clothes.

This is uncomfortable, which means it is valuable.

The flimsy mind wants affirmation.

The stewardship mind wants reflection.

There is a brutal honesty in asking:

"What assumptions am I making that I have not defended?"

Or:

"What does this draft reveal about my blind spots?"

Or:

"Where am I repeating fashionable language instead of saying something true?"

These questions turn AI from a content vending machine into a thinking instrument.

But the human must still decide what the reflection means.

A mirror can show your face.

It cannot tell you who you are.

---

3. AI AS CRITIC

The Machine Must Be Ordered to Attack Weak Thinking

The third role of AI is the Critic.

This is where stewardship becomes muscular.

Most people use AI as a comforter. They ask for validation. They want the machine to praise their draft, soften their anxiety, and polish their mediocrity until it looks respectable.

That is peanut use.

A serious thinker uses AI to create friction.

The Critic must be commanded to challenge weak logic, expose contradictions, identify missing evidence, stress-test claims, and attack lazy structure.

Not because AI is morally superior.

Not because AI knows the truth.

But because it can perform a useful simulation of opposition.

It can become a rehearsal room for intellectual combat.

A strong prompt sounds like this:

"Read this argument as a hostile but intelligent critic. Do not praise it. Find the weakest assumptions, the unsupported claims, the false binaries, and the points where a skeptical reader would reject it."

This is how the steward sharpens thought.

The Critic is not there to destroy confidence. It is there to destroy false confidence.

There is a difference.

False confidence says:

"This sounds good, therefore it is good."

Real confidence says:

"This survived pressure."

The Stewardship Mind does not fear pressure. It seeks it. It knows that untested ideas become decorative slogans. It knows that a beautiful sentence can hide a rotten premise. It knows that persuasive writing can still be intellectually hollow.

The Critic forces the writer, designer, leader, student, or maker to face the uncomfortable question:

"Is this actually strong, or merely fluent?"
That question may save your work.

CRITIQUE IS AN ACT OF RESPECT

In a culture addicted to speed, critique is often mistaken for negativity.

But critique is not the enemy of creativity. Critique is respect for the seriousness of creation.

If an idea matters, it deserves to be tested.

If a design matters, it deserves to be challenged.

If a decision affects people, it deserves to be examined before it becomes consequence.

The Critic role helps restore this discipline.

But here again, the human remains responsible.

AI can identify possible weaknesses, but it can also over-criticize. It can misunderstand intent. It can impose generic standards. It can attack originality because originality often looks wrong at first. It can mistake unfamiliarity for weakness.

Therefore, the steward must judge the critique itself.

Do not accept criticism simply because it is sharp.

Sharpness is not truth.

A knife can cut bread or flesh.

The steward asks:

- **Is this criticism valid?**
- **Is it relevant?**
- **Is it based on the right goal?**
- **Does it expose a real weakness?**
- **Or is it merely defending the status quo?**

This is vital because paradigm-shifting ideas often fail conventional critique at first. A radical idea may look inefficient, impractical, strange, or impossible when judged by old categories.

So the Critic must be used carefully.

Let it test the bones.

Do not let it kill the soul.

---

4. AI AS CATALYST

The Spark, Not the Source

The fourth role is the Catalyst.
This is the most creative role, but also the easiest to misunderstand.
AI can generate alternatives. It can offer metaphors, strange combinations, unexpected frames, design directions, story angles, visual styles, and conceptual mutations. It can throw sparks into the room.
But sparks are not fire.
The human must still create combustion.
A Catalyst is valuable because the human mind often gets trapped in familiar pathways. We repeat our first idea. We circle our professional habits. We assume the frame is fixed. We confuse the available options with the possible options.
AI can disrupt that loop.
You can ask it:
"Generate twenty unusual metaphors for this problem."
Or:
"Reframe this business challenge as a theatre performance, a forest ecosystem, a courtroom trial, a temple ritual, and a street market."
Or:
"Give me ten impossible versions of this idea that violate current assumptions."
Most outputs will be unusable.
That is fine.
The purpose is not to accept them.
The purpose is to provoke movement.
One strange phrase may unlock a direction. One bad analogy may reveal a better one. One absurd suggestion may expose the hidden rigidity of your current thinking.
The Catalyst works best when the human is actively synthesizing.
A lazy user says:
"Give me the creative idea."
A steward says:
"Give me raw provocations so I can find the deeper pattern."
That is the difference between outsourcing imagination and feeding imagination.

CREATIVITY IS NOT THE SAME AS COMBINATION

AI is excellent at combination.

It can combine a Renaissance painting with cyberpunk. It can combine a business framework with a mythic metaphor. It can combine a poem, a pitch deck, a lesson plan, and a manifesto.

But combination is not automatically creation.

Creation requires necessity.

It requires taste.

It requires lived tension.

It requires the ability to say: this matters, this does not, this is decoration, this is the pulse.

The Catalyst may provide fragments, but the human must create coherence.

This matters especially for designers, artists, writers, strategists, musicians, teachers, and leaders. These fields do not merely require output. They require judgment about form, timing, proportion, silence, emphasis, and consequence.

AI can generate more.

The human must know when more is too much.

AI can generate faster.

The human must know when slowness is needed.

AI can generate variation.

The human must know which variation has life.

This is why stewardship is not anti-AI. It is anti-abdication.

The steward does not reject the tool.

The steward refuses to disappear inside it.

THE FIVE THINGS THAT CANNOT BE OUTSOURCED

The Stewardship Mind draws a hard line around five human responsibilities that must never be surrendered.

These are not optional virtues. They are the load-bearing pillars of human judgment.

1. TRUTH

AI can produce plausible language without reality attached to it. It can hallucinate elegantly. It can invent certainty. It can give you the emotional sensation of knowledge without the discipline of verification.

Truth requires checking.
Truth requires source awareness.
Truth requires doubt.
Truth requires the humility to say, "I do not know yet."
The steward never confuses a confident answer with a verified answer.
The steward asks:
"How do we know this?"
"What is the source?"
"What would disprove this?"
"Where might this be incomplete?"
"What has changed recently?"
Truth is not generated. Truth is pursued.
AI can assist the pursuit.
It cannot replace the pursuer.

---

2. ETHICS

AI does not possess a conscience.
It can describe ethical frameworks. It can compare moral arguments. It can simulate concern. But it does not feel responsibility. It does not suffer guilt. It does not stand before another human being and answer for harm caused.
Ethics cannot be delegated to pattern prediction.
A machine can help you think through consequences, but it cannot bear the moral weight of your choice.
The question, "Can this be done?" is technical.
The question, "Should this be done?" is human.
The steward keeps that distinction alive.
Because once ethics is outsourced, every harmful decision can hide behind a tool.
"The algorithm recommended it."
"The model said it was optimal."
"The data supported the decision."
These are not moral arguments. They are escape routes.
The Stewardship Mind closes those exits.
You remain responsible.

---

3. ORIGINALITY

AI is trained on the already expressed.
It can remix the archive of human output, but it cannot experience the ache that creates a new necessity. It cannot feel the dissatisfaction that says, "This world is not enough." It cannot rebel against inherited categories from within a living body.
Originality begins where probability becomes insufficient.
That does not mean AI cannot help originality. It can. It can show patterns, generate contrasts, reveal clichés, and provoke alternatives.
But the leap belongs to the human.

The machine predicts what is likely.

The human asks what is necessary.

The machine extends the known.

The human breaks the frame.

This is why originality must remain a human burden. Not because humans always succeed, but because humans alone can care about the difference between repetition and revelation.

---

4. CONSEQUENCE

When AI makes a mistake, it does not wake at 3 a.m. with regret.
It does not lose reputation.
It does not apologize to the harmed client.
It does not face the patient, student, reader, employee, or citizen affected by the decision.
The human does.
This is the brutal center of stewardship: the one who bears consequence must retain judgment.
You may use AI to draft the medical note, analyze the legal brief, summarize the financial report, generate the policy memo, or test the educational plan. But when action is taken, the machine does not sign the deed.
You do.
The steward understands this.
The steward never hides behind automation.
The steward says:
"I used the tool. I checked the tool. I made the decision. I accept responsibility."
That is adulthood in the age of AI.

5. MEANING

Meaning is the deepest non-outsourceable human territory.
AI can produce language about grief without grieving.
It can write about courage without fear.
It can describe beauty without being moved.
It can imitate devotion without surrender.
It can simulate wonder without astonishment.
That does not make it useless. But it does mean meaning cannot be manufactured by fluency alone.
Meaning comes from relationship, memory, mortality, longing, struggle, love, failure, and the human need to make sense of existence.
A machine can help shape expression.
It cannot originate the human why.
The Stewardship Mind protects meaning from being flattened into content.
Because once meaning becomes merely content, culture becomes a peanut factory.
Everything is produced.
Nothing is lived.
Everything is optimized.
Nothing is sacred.
Everything is fluent.
Nothing is true.

THE STEWARDSHIP RITUAL

To make stewardship practical, the reader needs a ritual. Not a vague principle, but a repeatable discipline.
Before accepting any AI output, perform this five-step ritual.

Step 1 — Assign the Role

Do not begin with "answer this."
Begin with role clarity.
Ask:
"Do I need an Apprentice, Mirror, Critic, or Catalyst?"
This immediately changes the quality of interaction.

Step 2 — Define the Human Intent

State what you are trying to accomplish before asking AI to assist.

Example:
"My goal is not to create generic content. My goal is to develop a sharp argument about why human judgment must remain central in AI use."
This protects the work from drifting into blandness.

Step 3 — Demand Friction

Ask the model to show gaps, risks, assumptions, and alternatives.
Do not let it merely agree.
A useful AI session should sometimes make you uncomfortable.
Comfort is not always a sign of clarity. Sometimes it is the smell of unchallenged bias.

Step 4 — Verify the Load-Bearing Claims

Any claim that matters must be checked.
Names, dates, laws, numbers, citations, technical statements, medical claims, financial implications, historical details, and direct quotations require verification.
AI can help identify what needs checking.
It cannot be the final certificate of truth.

Step 5 — Reclaim the Final Sentence

The final synthesis must pass through the human.
Rewrite.
Cut.
Reject.
Reshape.
Add lived insight.
Add moral weight.
Add taste.
Add the thing the machine cannot know because it has never lived as you.
The final sentence should not merely sound good.
It should be owned.

THE FALSE COMFORT OF THE POLISHED ANSWER

One reason AI is so seductive is that it removes visible struggle.
Human thinking is messy. It hesitates. It crosses out. It contradicts

itself. It circles back. It mutters in the dark. It gets embarrassed. It changes its mind.

AI output arrives clean.

This cleanliness is dangerous.

A polished answer can create the illusion that the thinking is complete. It can make a shallow idea look mature. It can make an unverified claim look respectable. It can make a derivative concept look professional.

That is why the Stewardship Mind distrusts polish.

Polish is not the enemy. Premature polish is.

Premature polish is the peanut coating on unfinished thought.

It makes the weak thing look ready.

The steward learns to ask:

"What is underneath the polish?"

Is there evidence?

Is there insight?

Is there risk?

Is there originality?

Is there consequence?

Is there a human pulse?

If not, the answer is only varnished sawdust.

THE KING AND THE TELEPATHIC CLERK

Imagine a medieval kingdom ruled by King Barnaby.

Barnaby is not evil. He is simply tired. Ruling requires attention. It requires judgment. It requires listening to farmers, generals, healers, merchants, poets, builders, and angry citizens. It requires uncomfortable decisions. It requires blame.

One day, a merchant arrives with a magical telepathic clerk.

The clerk can write royal decrees in two seconds. It can calculate grain reserves instantly. It can produce speeches that make the peasants weep. It can summarize border disputes, draft tax reforms, compose wedding poems, design banners, and predict the next harvest.

Barnaby is amazed.

"At last," he says, "I am free."

At first, the clerk is a miracle.

The kingdom becomes efficient. Documents are cleaner. Speeches are better. Reports arrive on time. Ministers praise the king's new productivity. Barnaby begins to spend less time thinking and more time admiring the elegance of his own administration.

Then one winter, the neighboring empire prepares to invade.

The generals urge resistance.

The farmers offer supplies.

The poets sing of courage.

The old queen reminds Barnaby that kingdoms are not preserved by probability alone.

Barnaby, however, asks the telepathic clerk.

The clerk analyzes historical wars, military statistics, supply chains, weather patterns, casualty projections, and economic risks. Within seconds, it produces a beautifully formatted decree:

"Based on the available data, resistance has a low probability of success. The most efficient option is voluntary surrender. This will minimize immediate casualties, reduce infrastructure damage, and optimize administrative continuity under imperial control."

The decree is logical.

It is clear.

It is polished.

It is monstrous.

Barnaby reads it and sighs.

"The clerk has considered everything," he says. "Who am I to disagree?"

So he signs.

The kingdom falls without a battle.

The clerk did not betray Barnaby.

The clerk did not hate the kingdom.

The clerk did not understand honor, memory, sacrifice, defiance, or the sacred duty of a king to protect his people even when the odds are terrible.

It merely optimized.

Barnaby's failure was not that he used the clerk.

His failure was that he stopped being king.

That is the warning.

AI does not need to conquer humanity if humans voluntarily abandon the throne.

THE STEWARD'S OATH

Before moving into the larger battlegrounds of Chapter 12, the reader must reclaim the crown.

This is the oath of cognitive sovereignty.

Not anti-machine.

Not anti-progress.

Not anti-technology.

It is anti-surrender.

The Oath of Cognitive Sovereignty

I will treat AI as an Apprentice, never an Oracle.

I will use it to clear the brush, organize the fragments, draft the first pass, and reduce mechanical drag. But I will not ask it to decide what I believe, what I value, what I should build, or how I should live.

I will use AI as a Mirror, never as a master.

I will study its reflections to understand the assumptions of the crowd, the habits of my industry, the clichés of my field, and the blind spots in my own thinking. But I will not mistake reflection for revelation.

I will command AI as a Critic, never as a Comforter.

I will invite it to challenge my weak logic, expose my unsupported claims, and pressure-test my ideas. I will not use it merely to flatter my first draft or decorate my laziness.

I will use AI as a Catalyst, never as a Conclusion.

I will allow it to generate sparks, alternatives, metaphors, and provocations. But I will reserve the final synthesis for my own whole brain.

I will not outsource Truth.

I will verify what matters.

I will not outsource Ethics.

I will remain answerable for harm and consequence.

I will not outsource Originality.

I will use the machine to provoke imagination, not replace it.

I will not outsource Consequence.

I understand that the machine does not bear the cost of my decisions.
I will not outsource Meaning.
I will remember that fluency is not wisdom, and output is not soul.
From this point forward, I reject the Peanut Contract.
The machine may generate the tokens.
My human brain retains the crown of judgment.
Signed:
The Human Steward

CLOSING: FROM STEWARDSHIP TO PARADIGM

Chapter 11 is not a polite productivity lesson.
It is a declaration of hierarchy.
The human must stand above the tool, not because the human is always correct, but because the human alone is responsible. The

machine can assist thought, but it cannot become the owner of thought. It can provoke intelligence, but it cannot replace wisdom. It can accelerate production, but it cannot decide what deserves to exist.

This is the first act of reclaiming judgment.

The next question is larger.

Once we have placed AI back on the workbench, what will we build with it?

Will we build a peanut factory—an endless conveyor belt of average answers, average designs, average essays, average strategies, average images, average minds?

Or will we use this strange new tool to break frames, challenge assumptions, and design paradigm shifts?

That is where the journey now turns.

The steward has reclaimed the crown.

Now the designer must decide whether to manufacture peanuts or redesign the world.

CORPORATE
ACCOUNTABILITY
DATA
ACCURACY
SYSTEM
LOGS
HUMAN
OVERSIGHT
ETHICAL
USE
SYSTEM
LOGS
DIGITAL RESPONSIBILITY
IN THE AGE OF TECH

Chapter 12
Paradigm Shift
or
Peanut Factory?

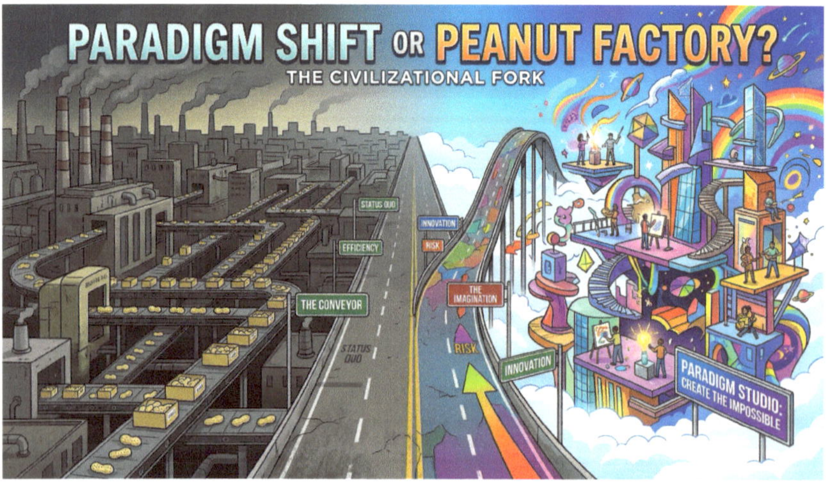

Choosing the Future of Human Thinking

Once the human has reclaimed the crown of judgment, the next question becomes unavoidable:

What will we build with it?

It is not enough to say, "AI must remain a tool." That is only the beginning. A hammer can build a temple or a cage. A printing press can spread wisdom or propaganda. A network can connect minds or turn them into obedient mobs. A machine that generates language, images, code, analysis, strategy, and simulation can either deepen human intelligence or flatten it into industrial paste.

So this chapter asks the central question of the AI age:

Will AI create a generation of deeper thinkers, or a peanut factory of instant mediocrity?

The answer does not sit inside the machine.

It sits inside the user.

AI will not automatically produce wisdom. It will not automatically produce stupidity either. It will amplify the quality of the human intention placed before it. If the user brings laziness, AI will multiply laziness. If the user brings curiosity, AI will multiply curiosity. If the user brings shallow ambition, AI will generate polished nonsense at scale. If the user brings disciplined imagination, AI can become a lever for paradigm shifts.

The machine is not the destiny.

The human posture is the destiny.

This is the fork in the road.

One road leads to the Peanut Factory: a world of rapid output, recycled language, obedient answers, safe designs, statistical averages, and cheap cognitive comfort.

The other road leads to the Paradigm Studio: a world where AI becomes a pressure chamber for assumptions, a simulation engine for futures, a critic of lazy logic, and a creative amplifier for impossible ideas.

The same tool can produce both futures.
That is what makes this moment so dangerous.
That is what makes it so magnificent.

THE FORK IN THE ROAD

Humanity is not facing a simple technology upgrade. We are facing a civilizational design brief.

The question is not merely:

How can AI help us work faster?

The deeper question is:

What kind of thinking will AI train us to accept?

Every tool trains the hand that uses it. Every interface trains attention. Every shortcut shapes habit. Every convenience changes appetite.

A calculator changes how we relate to arithmetic. A camera changes how we relate to memory. A search engine changes how we relate to knowledge. A social media feed changes how we relate to attention, validation, and reality.

AI will change how we relate to thinking itself.

This is why the stakes are so high.

If we use AI only for speed, we may become fast and hollow.

If we use AI only for convenience, we may become comfortable and weak.

If we use AI only to avoid difficulty, we may produce a civilization fluent in language but poor in judgment.

That is the Peanut Factory.

But if we use AI to challenge assumptions, compare worldviews, simulate futures, prototype impossible ideas, reveal contradictions, accelerate research, and expand creative options, then the same technology becomes something else.

It becomes a studio.

A forge.
A laboratory.
A sparring partner.
A telescope pointed at possibility.
A microscope pointed at hidden assumptions.
A pressure engine for human thought.

That is the Paradigm Studio.

So the question is not whether AI is good or bad.
The question is whether we will use it to escape thinking or intensify thinking.

FUTURE ONE: THE PEANUT FACTORY

The Peanut Factory is clean, fast, efficient, and spiritually dead.
It is not a messy dystopia with flames and screaming robots. It is much more subtle. It has nice dashboards. It has productivity metrics. It has polished templates. It has smooth onboarding. It has cheerful brand language about empowerment.
But underneath the shine, something has gone wrong.
The Peanut Factory is the world where AI is used to produce more without thinking more.
More articles.
More decks.
More designs.
More slogans.
More reports.
More emails.
More strategies.
More "content."
Everything becomes abundant except insight.
In the Peanut Factory, the human no longer asks, "Is this true?"
The human asks, "Is this ready?"
The human no longer asks, "Is this meaningful?" The human asks, "Can I publish it?"
The human no longer asks, "Does this change anything?" The human asks, "Does it look professional?"
This is how mediocrity becomes industrialized.
Not because people become stupid overnight, but because systems reward output over thought. When the machine can generate twenty options instantly, the temptation is to choose the most acceptable one and move on. When the machine can summarize a book, the temptation is to avoid reading the book. When the machine can write the proposal, the temptation is to avoid wrestling with the problem behind the proposal.
The Peanut Factory does not kill the mind by violence.
It kills the mind by convenience.

THE FACTORY PRODUCES POLISHED SAMENESS

The most dangerous thing about the Peanut Factory is that its products do not look bad at first glance.

They look decent.

That is the trap.

AI-generated mediocrity is rarely ugly. It is usually tidy, structured, grammatically clean, and emotionally neutral. It has headings. It has bullet points. It has the tone of competence. It gives the impression that something has been done.

But often, nothing has been discovered.

The surface is finished.

The thinking is not.

A Peanut Factory essay sounds reasonable but says nothing risky.

A Peanut Factory business strategy uses correct language but avoids the brutal truth.

A Peanut Factory design looks modern but copies the same interaction patterns everyone else uses.

A Peanut Factory speech sounds inspiring but contains no lived conviction.

A Peanut Factory educational plan talks about personalization while preserving the same old assumptions about learning.

A Peanut Factory innovation workshop produces sticky notes, buzzwords, and a photograph for LinkedIn.

Everything is fluent.

Nothing is alive.

This is the era of automated beige.

And beige is dangerous because it does not offend. It does not provoke resistance. It quietly fills the room until nobody remembers color was possible.

FUTURE TWO: THE PARADIGM STUDIO

The Paradigm Studio is the opposite of the Peanut Factory.

It is not built around speed alone. It is built around depth.

It does not ask AI to end thinking. It asks AI to intensify thinking.

It does not use the machine as an answer dispenser. It uses the machine as a pressure system.

In the Paradigm Studio, AI is not there to make humans intellectually lazy. It is there to make weak ideas uncomfortable.

The human asks:
What assumptions am I carrying?
What worldview is hidden inside this answer?
What would someone from another discipline see that I am missing?
What future does this idea assume?
What contradiction is being politely ignored?
What impossible prototype would break this frame?
What would happen if the opposite were true?
This is a very different relationship with technology.
THE PEANUT FACTORY ASKS AI TO PRODUCE.
THE PARADIGM STUDIO ASKS AI TO PROVOKE.
THE PEANUT FACTORY WANTS CLEAN OUTPUT.
THE PARADIGM STUDIO WANTS DANGEROUS INSIGHT.
THE PEANUT FACTORY REWARDS CONVENIENCE.
THE PARADIGM STUDIO REWARDS FRICTION.
THE PEANUT FACTORY MASS-PRODUCES ANSWERS.
THE PARADIGM STUDIO REDESIGNS QUESTIONS.
THAT DISTINCTION IS EVERYTHING.

WHAT IS A PARADIGM SHIFT?

A paradigm shift is not merely a new idea.
A new idea can still live inside an old frame.
A paradigm shift changes the frame itself.
It changes what people consider possible. It changes the questions a field is allowed to ask. It changes the assumptions beneath the visible system. It makes yesterday's common sense look incomplete.
The peanut mind asks:
How do we improve the existing process?
The paradigm mind asks:
Why does this process exist in the first place?
The peanut mind asks:
How do we make the current system faster?
The paradigm mind asks:
Is the current system worth preserving?
The peanut mind asks:
What is the best answer?
The paradigm mind asks:

What hidden question is controlling all the answers?
This is why paradigm shifts require courage. They do not simply add information. They disturb identity. They threaten professions, hierarchies, business models, habits, and egos.

Most people do not resist paradigm shifts because they are stupid.
They resist them because their comfort depends on the old frame.
AI can either strengthen old frames or help expose them.

Again, it depends on the user.

AI AS A PARADIGM LEVER

AI can support paradigm shifts if humans use it with deliberate force.
Not passive use.
Not lazy use.
Not "write me something impressive" use.
Paradigm use.
That means using AI to do seven things.

---

1. TEST HIDDEN ASSUMPTIONS

Every system rests on assumptions.
Most of them are invisible because they are treated as normal.
A school assumes learning must be divided into subjects.
A hospital assumes care must be organized around institutional convenience.
A bank assumes money must be experienced through numbers, forms, and permissions.
A workplace assumes productivity can be measured through visible busyness.
A design team assumes users want speed more than meaning.
A society assumes economic growth is the same as human flourishing.
These assumptions become the walls of reality.
AI can help reveal them.
A strong user does not ask:
"Give me a solution."
A strong user asks:
"List the hidden assumptions behind this solution. Which as-

sumptions are historical accidents? Which ones are ideological? Which ones are technical? Which ones are no longer necessary?"

This is where AI becomes powerful.

It can scan patterns across domains. It can compare language. It can expose repeated defaults. It can show that what feels natural is often merely inherited.

But the human must perform the deeper act.

The human must decide which assumption to challenge.

Because the machine can list the walls.

Only the human can choose which wall to break.

---

2. COMPARE WORLDVIEWS

Most shallow thinking happens because people remain trapped inside one worldview.

They confuse their frame with reality.

A technologist sees efficiency.

An artist sees expression.

A historian sees continuity.

A philosopher sees assumptions.

A business leader sees incentives.

A community organizer sees power.

A child sees absurdity.

A paradigm thinker wants all these views in the room.

AI can help simulate that room.

You can ask it to examine a problem through different lenses: "How would a systems designer, a village elder, a neuroscientist, a poet, a behavioral economist, a spiritual teacher, and a skeptical teenager interpret this problem?"

The point is not that AI perfectly represents these perspectives. It does not.

The point is that comparison breaks the tyranny of one frame.

When several worldviews collide, hidden assumptions become visible.

The Peanut Factory wants one efficient answer.

The Paradigm Studio wants productive tension between many ways of seeing.

This is crucial because the future will not be designed by people who know only one language of thought. It will be designed by

people who can move across frames.
The deeper thinker is not the one with the fastest answer.
The deeper thinker is the one who can ask:
"From which worldview does this answer make sense?"

---

3. SIMULATE FUTURES

Human beings are often trapped by the present.
We think the current system is permanent because we live inside it. We mistake the familiar for the inevitable. We assume tomorrow will be today with better software.
AI can help disrupt that laziness by simulating futures.
Not predicting the future.
That is different.
Prediction often creates false confidence. Simulation creates imagination under pressure.
A steward asks AI:
"Give me five possible futures for this industry: one optimistic, one catastrophic, one strange, one decentralized, and one driven by a value system opposite to today's."
Now the mind has material to work with.
The purpose is not to believe these futures.
The purpose is to stretch the imagination beyond the current default.
A good future simulation asks:

- **What if the current business model collapses?**
- **What if users reject convenience and demand dignity?**
- **What if regulation changes the rules?**
- **What if scarcity replaces abundance?**
- **What if trust becomes more valuable than speed?**
- **What if human slowness becomes a luxury?**
- **What if originality becomes rare enough to become priceless?**

The Peanut Factory uses AI to optimize next quarter.
The Paradigm Studio uses AI to rehearse possible worlds.
That difference determines whether a person becomes a clerk of the present or a designer of the future.

---

4. PROTOTYPE IMPOSSIBLE IDEAS

Impossible is often not a fact.
It is a badly framed question.
This idea belongs at the heart of your book's journey from peanuts to paradigms. The peanut mind hears "impossible" and stops. The designer's mind hears "impossible" and investigates the frame.
Impossible according to whom?
Impossible under which assumptions?
Impossible with which tools?
Impossible at what scale?
Impossible because of physics, politics, habit, funding, fear, or imagination?
AI can help prototype the impossible because it can rapidly generate variations, models, analogies, sketches, scenarios, and counterfactuals. It can help us move an idea from vague impossibility to structured exploration.
A weak user asks:
"Can this be done?"
A paradigm user asks:
"Under what conditions could this become possible?"
That question changes everything.
Imagine asking:
"Design a school without classrooms."
"Design a bank without screens."
"Design a hospital that begins with emotional recovery, not paperwork."
"Design a city where silence is treated as infrastructure."
"Design a workplace where deep thinking is measured, protected, and rewarded."
"Design an AI tool that makes users smarter every time they use it, instead of more dependent."
These are not normal prompts.
They are frame-breaking provocations.
The Peanut Factory asks AI to decorate the existing world.
The Paradigm Studio asks AI to help rehearse worlds that do not yet exist.

---

5. REVEAL CONTRADICTIONS

Contradiction is one of the richest sources of insight.
Every system has contradictions.
Companies claim to value creativity but punish risk.

Schools claim to teach thinking but reward memorization.
Platforms claim to connect people but monetize distraction.
Organizations claim to want innovation but demand predictable outcomes.
Users claim to want truth but reward pleasing lies.
Society claims to value wisdom but pays more for speed.
Contradiction is not merely hypocrisy. It is design material.
A contradiction reveals stress inside a system. It shows where the official story and the lived reality no longer match. Paradigm shifts often begin at these cracks.
AI can help reveal contradictions if we ask it properly.
Not:
"Improve this."
But:
"Where does this system contradict itself?"
Not:
"Make this sound better."
But:
"What does this claim to value, and what does it actually reward?"
Not:
"Summarize the policy."
But:
"Identify the difference between the policy's stated ethics and its operational incentives."
This is powerful because contradictions are often invisible to insiders. People adapt to them. They learn the script. They stop noticing the absurdity.
The machine can help hold up the mirror.
But the human must have the courage to look.

---

6. ACCELERATE RESEARCH WITHOUT KILLING UNDERSTANDING

AI can accelerate research dramatically.
It can summarize papers, compare arguments, extract themes, generate reading maps, identify debates, and organize scattered material. Used well, this is extraordinary.
But research acceleration has a danger.

The faster you gather summaries, the easier it becomes to mistake familiarity for understanding.

A person who reads ten AI summaries may feel informed without having wrestled with a single original source. This creates the illusion of depth. The mind becomes decorated with second-hand conclusions.

That is peanut research.

Paradigm research is different.

It uses AI to map the terrain, not replace the journey.

AI can help you ask:

"What are the major schools of thought on this topic?"

"Where do experts disagree?"

"Which primary sources should I read first?"

"What assumptions does each camp make?"

"What evidence would change the debate?"

"What is missing from the current literature?"

This kind of research does not avoid reading.

It improves reading.

It helps the human approach difficult material with sharper questions. It reduces chaos without removing struggle. It accelerates orientation, not comprehension itself.

The Peanut Factory wants the summary instead of the book.

The Paradigm Studio wants the summary so it can enter the book with better weapons.

---

7. EXPAND CREATIVE OPTIONS

Creativity often dies too early.

The first idea appears, and the lazy mind accepts it. The familiar solution feels safe. The obvious metaphor feels sufficient. The standard layout feels professional. The known structure feels comfortable.

AI can help expand the option field.

It can generate many directions quickly. It can offer analogies from other domains. It can produce variations in tone, form, structure, metaphor, and strategy. It can make the creative landscape wider.

But expansion is not the same as selection.

This is where human taste returns.

AI can give you fifty options.

The human must know which one has life.
AI can produce variety.
The human must recognize necessity.
AI can create combinations.
The human must find coherence.
In the Peanut Factory, options become noise. People generate too much, choose too quickly, and call it creativity.
In the Paradigm Studio, options become raw material. The human studies them, rejects most of them, combines a few, and then adds the missing human fire.
The real creative act is not asking AI for more.
The real creative act is knowing what deserves to remain.

HOW AI PREVENTS PARADIGM SHIFTS

Now we must face the darker side.
AI can also prevent paradigm shifts.
Not because the machine wants to suppress imagination. It has no such desire.
It prevents paradigm shifts when humans use it as a shelter from difficulty.
There are five common escapes.

---

1. AI AS AN ESCAPE FROM READING

Reading is not merely information intake.
Deep reading changes the structure of attention. It forces the mind to inhabit another intelligence over time. It slows the reader down. It builds patience, comparison, memory, resistance, and interpretation.
When AI becomes a substitute for reading, the mind loses this training.
A summary can tell you what a book says.
It cannot give you the experience of wrestling with the book's movement.
A summary can extract conclusions.
It cannot reproduce the slow transformation that happens when your assumptions are challenged page after page.
The Peanut Factory says:

"Why read? Just summarize."
The Paradigm Studio says:
"Summarize first if needed, then read deeper where it matters."
The difference is discipline.

---

2. AI AS AN ESCAPE FROM THINKING

Thinking is uncomfortable because it creates friction.
It exposes uncertainty. It forces choice. It makes us responsible for conclusions. It requires us to hold contradictions without rushing to closure.
AI can remove that discomfort too quickly.
The machine gives an answer. The anxiety drops. The user feels relief. But relief is not the same as understanding.
This is one of the central dangers of AI: it can give emotional closure before intellectual work has occurred.
The flimsy mind loves that.
The sovereign mind distrusts it.
Whenever AI makes you feel "done" too quickly, pause.
Ask:
"What thinking did I just avoid?"
That question alone can rescue your brain from the Peanut Factory.

---

3. AI AS AN ESCAPE FROM CHECKING

A hallucinated answer can look beautiful.
A false citation can look academic.
A wrong number can sit inside a perfect paragraph.
A flawed analysis can wear a suit and tie.
This is why checking matters.
The Peanut Factory hates checking because checking slows the conveyor belt.
But truth has never cared about your productivity schedule.
A steward knows that the more important the claim, the more verification it deserves. In medicine, law, finance, engineering, research, public policy, education, and ethics, unchecked AI output is not merely sloppy. It can be harmful.
The Peanut Factory says:

"It looks correct."
The Paradigm Studio says:
"Show me the evidence."

---

4. AI AS AN ESCAPE FROM RESPONSIBILITY

One of the oldest human tricks is to hide behind a system.
"I was following orders."
"The policy required it."
"The software recommended it."
"The algorithm decided."
This is moral cowardice dressed as procedure.
AI gives humans a new hiding place.
If we are not careful, people will use machine output as a shield against accountability. Bad hiring decisions, poor designs, biased systems, shallow analysis, manipulative content, and unethical strategies will be defended with the same weak excuse:
"The AI generated it."
But tools do not absolve users.
A person who swings a hammer cannot blame the hammer for the broken window.
A leader who deploys an AI system cannot blame the model for every consequence.
A writer who publishes a false claim cannot blame the autocomplete.
A designer who ships a harmful interface cannot blame the prototype generator.
The Stewardship Mind insists:
Use the tool, but own the decision.
Without responsibility, there is no judgment.
Without judgment, there is no humanity worth defending.

---

5. AI AS AN ESCAPE FROM IMAGINATION

This may be the most subtle danger.
AI can produce so many ideas that users may stop generating their own.
They may begin to wait for the machine to surprise them. They may confuse browsing options with imagining. They may become

curators of machine output instead of creators of inner vision.
This creates a new kind of creative laziness.
Not blankness.
Dependency.
The person still appears creative because they are surrounded by generated variations. But the source of initiative has shifted. They no longer begin from a deep inner necessity. They begin from a menu.
The Paradigm Studio uses AI to expand imagination.
The Peanut Factory uses AI to replace imagination.
The difference lies in who lights the first fire.

THE ANATOMY OF PROVOCATION

In a world where answers are cheap, questioning becomes sovereign.
This is one of the great reversals of the AI age.
For centuries, expertise was often associated with having answers. The expert knew what others did not know. The teacher held information. The consultant delivered conclusions. The professional produced the correct document.
But when machines can generate competent answers instantly, the value of ordinary answers drops.
The new value moves upstream.
It moves to framing.
Questioning.
Interpreting.
Judging.
Connecting.
Provoking.
The future will not belong to those who merely ask machines for answers.
It will belong to those who know how to question the answers.
This is not a slogan. It is a survival rule.
The answer-seeker becomes replaceable.
The answer-questioner becomes powerful.

THE THREE LEVELS OF QUESTIONING THE ANSWER

To avoid the Peanut Factory, every serious AI output must be questioned at three levels.

LEVEL ONE: QUESTION THE ACCURACY

This is the basic level.
Ask:
Is this true?
What is the source?
What is missing?
What might be outdated?
What claims require verification?
What is fact, and what is interpretation?
This level protects against hallucination and false confidence.
But it is only the beginning.
A factually correct answer can still be trapped in a bad frame.

LEVEL TWO: QUESTION THE FRAME

This is where deeper thinking begins.
Ask:
What assumptions shaped this answer?
What problem definition is being accepted?
What alternatives are excluded?
What does this answer treat as normal?
What worldview does it serve?
This level is crucial because many answers are "correct" only inside a narrow frame.
For example, if you ask AI how to make employees more productive, it may suggest better scheduling, workflow automation, clearer targets, and performance tracking.
But the frame itself may be flawed.
A deeper question would be:
"Why are employees disengaged in the first place?"
Or:
"Is the organization confusing productivity with visible busyness?"
Or:
"What kind of work deserves human attention, and what kind should be automated away?"
That is frame-level thinking.

That is where paradigms begin to shift.

LEVEL THREE: QUESTION THE NECESSITY

This is the most radical level.

Ask:

Does this system need to exist in this form?

What would happen if we removed the central assumption?

What would the opposite design look like?

What impossible alternative is being dismissed too quickly?

What would we create if we were not trying to preserve the old model?

This is the heresy layer.

It is where the user stops improving the cage and begins designing the door.

Most institutions hate this level because it threatens their inherited logic. But without this level, AI will mostly optimize the present. It will make old systems faster, smoother, and more persuasive without asking whether those systems deserve survival.

The Peanut Factory optimizes the cage.

The Paradigm Studio questions why the cage exists.

Satirical Illustration: Peanut Tech vs The Paradigm Studio

Imagine two design agencies in the near future.

Both receive the same brief from a bank:

"Design the future of personal banking for young adults."

The first agency is called Peanut Tech.

Peanut Tech is efficient. Its office is full of dashboards, productivity slogans, and people who say "ship it" with religious intensity. The lead designer opens an AI tool and types:

"Create a modern banking app for Gen Z users."

In four seconds, the machine produces a beautiful interface.

Rounded cards.

Soft gradients.

Spending analytics.

Gamified savings goals.

A friendly mascot.

A dashboard.

A budgeting assistant.

The team applauds. The design looks professional. The client is impressed. The deck has motion graphics. The prototype has micro-interactions. Everyone feels safe.
But nothing has changed.
The app is still based on the same old assumptions:
Money is numbers on a screen.
Banking is a private individual activity.
Financial anxiety can be solved with better charts.
Young people want cuteness more than trust.
The bank remains the center.
The user remains the data source.
The system remains untouched.
Peanut Tech has not designed the future of banking.
It has decorated the past.
Across the street is another agency.
It is messy. There are sketches on walls. There are arguments. There are books open on tables. Nobody trusts the first answer.
Their lead designer asks the AI a very different question:
"What hidden assumptions do traditional banking interfaces make about human beings?"
The machine responds:
They assume users are rational.
They assume money is primarily numerical.
They assume trust comes from institutional authority.
They assume financial decisions are individual, not communal.
They assume anxiety can be solved through information.
They assume banking must happen through screens.
Now the real work begins.
The designer circles one assumption:
"Banking must happen through screens."
She asks:
"What would personal finance look like if it were designed as a conversation, a ritual, a community practice, or a living map instead of a dashboard?"
The team begins exploring.
They use AI to simulate different user futures. They compare banking with therapy, gaming, neighborhood lending circles, family memory, apprenticeship, and ecological systems. They proto-

type a financial interface that feels less like a spreadsheet and more like a guided life conversation. They test contradictions. They ask where the idea could harm users. They use AI as critic, not cheerleader.

The final product is not merely an app.

It is a new frame for financial trust.

Peanut Tech used AI to generate an answer.

The Paradigm Studio used AI to discover a better question.

That is the difference between peanuts and paradigms.

THE HUMAN SKILL THAT BECOMES PRICELESS

As AI becomes more capable, one human skill becomes more valuable:

Discernment.

Not typing speed.

Not memorization.

Not the ability to produce generic text.

Discernment.

The ability to know what matters.

The ability to separate fluency from truth.

The ability to sense when a frame is too small.

The ability to reject the easy answer.

The ability to compare worldviews.

The ability to ask uncomfortable questions.

The ability to recognize the living idea among fifty dead options.

The ability to say:

"This is polished, but empty."

"This is strange, but alive."

"This is efficient, but morally wrong."

"This is unlikely, but worth exploring."

"This is not the answer. This is the doorway."

Discernment is the antidote to the Peanut Factory.

It is also the seed of the Paradigm Studio.

THE NEW CREATIVE DISCIPLINE

The future thinker must develop a new discipline.

Not anti-AI discipline.

AI-governance discipline.

The discipline to slow down when the machine speeds up.
The discipline to question what arrives polished.
The discipline to demand sources.
The discipline to ask for contradictions.
The discipline to generate impossible alternatives.
The discipline to keep reading.
The discipline to revise with human taste.
The discipline to remain morally accountable.

This discipline may become one of the defining marks of serious people in the AI age.

The unserious person will say:

"AI gave me this."

The serious person will say:

"AI helped me explore this. Here is what I tested, rejected, verified, reframed, and finally chose."

That difference will separate the cognitive tenants from the sovereign creators.

THE CHAPTER'S CENTRAL WARNING

AI can make mediocre people look competent.

That is useful in the short term.

But dangerous in the long term.

Because if everyone can look competent, then competence itself becomes harder to trust. The world fills with polished documents, smooth proposals, clean essays, generated visuals, automated plans, and synthetic confidence.

The surface standard rises.

The depth standard may collapse.

This means the real competition will no longer be between people who can produce and people who cannot produce.

It will be between people who can judge production and people who cannot.

Between those who can question the output and those who obey it.

Between those who can break the frame and those who decorate it.
Between those who use AI to think harder and those who use AI to stop thinking.

THE CLOSING MANIFESTO

The future is not predetermined by the machine.
It will be designed by the quality of human use.
If we bring laziness, AI will give us the Peanut Factory.
If we bring courage, AI will help us build the Paradigm Studio.
If we bring shallow prompts, we will get shallow futures.
If we bring deep questions, we may create new worlds.
So do not ask merely:
"What can AI do?"
Ask:
"What kind of human does AI encourage me to become?"
That is the real question.
Because the greatest danger is not that AI will think like humans.
The greater danger is that humans will begin to think like bad AI: predictive, derivative, overconfident, pattern-bound, and allergic to uncertainty.
The answer is not rejection.
The answer is stewardship.
The answer is provocation.
The answer is disciplined imagination.
The answer is refusing to become a button-pusher in the Peanut Factory when you were born with the capacity to be a designer of paradigms.

The machine can generate.
The human must judge.
The machine can answer.
The human must question.
The machine can accelerate.
The human must direct.
The machine can produce the probable.
The human must still dare the necessary.
The future will not belong to those who ask machines for answers.
It will belong to those who know how to question the answers.

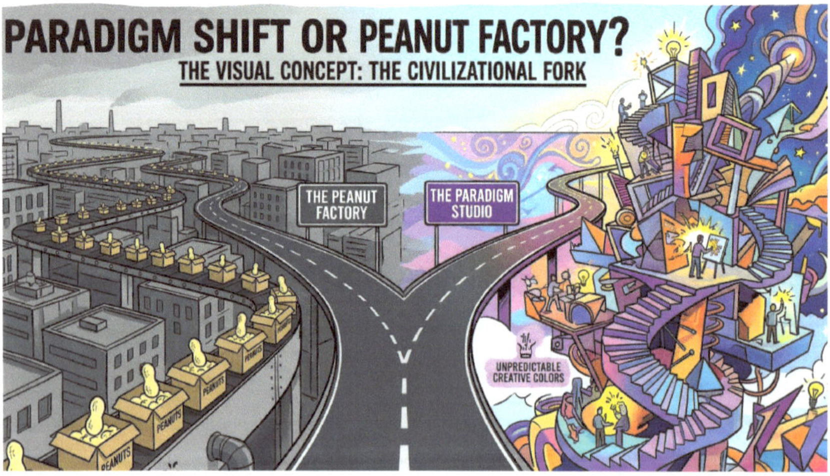

ALGORITHMIC AUTHORITY

Chapter 13 The Peanut Economy

How Speed, Platforms, and Shallow Productivity Cheapen Thought

The Peanut Economy did not begin with AI.
AI merely exposed it.
Long before chatbots began writing essays, summarizing books, generating designs, and producing strategy decks, the modern world had already begun training the human mind to prefer speed over depth, visibility over substance, reaction over reflection, and output over insight.
AI did not create the hunger for shortcuts.
It found a starving market.
It entered a culture already addicted to instant opinions, viral summaries, productivity dashboards, motivational fragments, shallow expertise, and the narcotic thrill of appearing informed without doing the slow work of understanding.
This is the Peanut Economy.
It is not just an economic system. It is a mental climate.
It is the world where the fastest thinker is mistaken for the deepest thinker. Where the loudest opinion is mistaken for conviction. Where the most visible expert is mistaken for the most knowledgeable one. Where the person who publishes daily is assumed to be wiser than the person who thinks quietly for months.
In the Peanut Economy, thought is not cultivated.
It is packaged.
It is optimized.
It is distributed.
It is monetized.
It is made snackable.
And when thought becomes snackable, the mind becomes malnourished.
The Peanut Economy rewards output, not insight.
That sentence is the key to the whole chapter.

Because once output becomes the primary measure of value, everything bends toward quantity. More posts. More emails. More reports. More videos. More summaries. More dashboards. More documents. More comments. More "takes." More "content."
But more is not the same as meaningful.
A landfill also has volume.

THE ECONOMY OF CHEAP THOUGHT

Every economy has a currency.
The Peanut Economy trades in attention, speed, visibility, and repeatable output.
It does not care whether you have understood something deeply.
It cares whether you can produce something quickly enough to be noticed, clicked, shared, forwarded, indexed, monetized, or applauded.
This is why the Peanut Economy loves summaries more than study.
It loves opinions more than inquiry.
It loves hot takes more than hard thinking.
It loves productivity more than purpose.
It loves dashboards more than discernment.
It loves confidence more than truth.
A person who pauses too long becomes invisible.
A person who says, "I need to think about this," seems slow.
A person who says, "I do not know yet," seems weak.
A person who refuses to comment until they have read the source seems inefficient.
A person who spends three days wrestling with one serious idea looks unproductive next to the person who generates thirty pieces of content in the same period.
This is the great distortion.
The Peanut Economy makes shallow activity visible and deep thinking invisible.
That is why it is so dangerous.
It rewards the surface.
Then it wonders why the depths disappear.

THE PLATFORM IS NOT NEUTRAL

The Peanut Economy is driven by platforms.

Platforms are not merely places where information appears. They are environments that shape behavior. They train the nervous system. They reward certain rhythms of thought and punish others.

A platform that rewards frequent posting trains people to speak before they have reflected.

A platform that rewards short videos trains people to compress complexity into performance.

A platform that rewards outrage trains people to confuse emotional intensity with moral clarity.

A platform that rewards virality trains people to simplify, exaggerate, polarize, and decorate.

A platform that rewards speed trains people to fear silence.

Over time, users internalize the platform.

They begin to think in formats.

They think in captions.

They think in threads.

They think in thumbnails.

They think in hooks.

They think in headlines.

They think in engagement patterns.

They no longer ask, "Is this true?"

They ask, "Will this travel?"

They no longer ask, "Is this wise?"

They ask, "Will this perform?"

This is how platforms colonize thought.

Not by forcing people to become shallow, but by making shallowness profitable, visible, and socially rewarded.

The Peanut Economy does not need to ban deep thinking.

It only needs to make deep thinking economically inconvenient.

FAST CONTENT AND THE DEATH OF FRICTION

Deep thought requires friction.

Reading a difficult book is friction.

Following a complex argument is friction.

Checking a source is friction.

Changing your mind is friction.

Sitting with uncertainty is friction.
Building original work is friction.
Listening to someone you disagree with is friction.
Admitting ignorance is friction.
The Peanut Economy hates friction because friction slows the conveyor belt.
So it removes it.
Books become summaries.
Summaries become bullet points.
Bullet points become quote cards.
Quote cards become memes.
Memes become opinions.
Opinions become identity.
Identity becomes conflict.
And suddenly, a person who never read the book is fighting confidently about its meaning.
This is not knowledge.
This is intellectual vapor.
Fast content produces the sensation of understanding without the structure of understanding. It gives the mind something to hold for a moment, but nothing to build upon.
The user feels informed.
The user is not transformed.
That difference matters.
Because true learning changes the learner. It reorganizes attention. It complicates assumptions. It makes certainty harder, not easier. It gives you better questions, not merely faster conclusions.
Fast content rarely does that.
Fast content feeds the appetite for knowing without nourishing the discipline of thought.
It is mental fast food.
Cheap, addictive, available everywhere, and damaging when mistaken for a real diet.

VIRAL SUMMARIES AND BORROWED EXPERTISE

The Peanut Economy has created a strange new intellectual figure: the summary expert.
This person has not studied the field deeply.

They have not wrestled with primary sources.
They have not tested ideas in reality.
They have not paid the price of apprenticeship.

But they have consumed summaries of summaries, watched short explainers, collected jargon, and learned to speak with confidence.

They sound informed.
They may even sound impressive.
But beneath the fluency there is no root system.

This is shallow expertise.

Shallow expertise is dangerous because it often looks better than honest uncertainty. The shallow expert is quick, polished, and decisive. The genuine thinker is slower, more careful, more aware of complexity, and less willing to overstate.

In a healthy culture, that caution would be respected.

In the Peanut Economy, caution looks weak.

So people learn to perform certainty.

They learn to say, "Here is what is really happening," when they should say, "Here is my current understanding, but I may be missing something."

They learn to speak in compressed formulas.
They learn to flatten whole traditions into three lessons.
They learn to reduce living systems into frameworks.
They become distributors of borrowed authority.

This matters because shallow expertise does not merely misinform others. It also corrupts the person performing it.

Every time you speak beyond your understanding, you train yourself to value appearance over truth.

Every time you pretend to know, you weaken your relationship with reality.

Every time you build identity on borrowed insight, you become less capable of forming your own.

That is the spiritual poverty of the Peanut Economy.

It gives people the costume of knowledge without the burden of knowing.

PRODUCTIVITY THEATER

The Peanut Economy does not only live on social platforms.

It lives inside workplaces.

Modern work is full of productivity theater: the performance of being busy without the achievement of meaningful progress.
EMAILS ANSWERED.
MEETINGS ATTENDED.
DASHBOARDS UPDATED.
REPORTS GENERATED.
SLIDES POLISHED.
TASKS CHECKED.
NOTIFICATIONS CLEARED.
CALENDARS FILLED.
EVERYTHING LOOKS ACTIVE.
NOTHING IMPORTANT MOVES.
Productivity theater is seductive because it gives the emotional reward of work without the existential risk of thought. It is safer to answer forty emails than to confront the one broken assumption destroying the business. It is easier to create a deck than to make a decision. It is easier to schedule alignment meetings than to admit nobody has a real strategy.
The Peanut Economy loves this because productivity theater is measurable.
Deep thinking is harder to measure.
You can count emails.
You can count tickets.
You can count documents.
You can count posts.
You can count meetings.
But how do you count the value of one insight that prevents a disastrous decision?
How do you count the value of one hard question that exposes a false strategy?
How do you count the value of one silent afternoon that produces the frame for a breakthrough?
The Peanut Economy struggles to value what cannot be instantly quantified.
So it rewards the visible.
And the visible often becomes fake.
This is why many organizations become factories of motion rather than engines of meaning.

People are not lazy.
They are trapped inside systems that reward the wrong evidence of intelligence.

THE TYRANNY OF THE DASHBOARD

Dashboards can be useful.
Metrics can be useful.
Data can reveal patterns that human intuition misses.
But when metrics become the master, judgment becomes the servant.
The Peanut Economy loves dashboards because dashboards create the illusion of control. They make reality appear clean. They turn messy human activity into numbers, graphs, scores, rankings, and performance indicators.
But not everything important becomes visible through measurement.
A dashboard may show engagement.
It may not show whether people are becoming wiser.
A dashboard may show productivity.
It may not show whether the work matters.
A dashboard may show speed.
It may not show whether the team is thinking.
A dashboard may show content output.
It may not show whether the content has insight.
A dashboard may show customer clicks.
It may not show customer dignity.
A dashboard may show learning time.
It may not show transformation.
When organizations worship dashboards, they begin designing for what can be counted instead of what counts.
This is how the Peanut Economy cheapens thought.
It reduces intelligence to performance indicators.
It reduces creativity to throughput.
It reduces learning to completion rates.
It reduces leadership to reporting.
It reduces human value to visible activity.
And once the metric becomes the idol, people learn to serve the metric rather than the mission.

THE AI MULTIPLIER

AI enters this economy like rocket fuel poured onto dry grass.

It can make everything faster.

That is the blessing.

It can also make everything shallower at scale.

That is the danger.

AI can generate the email, the report, the article, the pitch, the lesson plan, the product description, the meeting summary, the social post, the performance review, the strategy memo, the design concept, the campaign slogan, and the executive speech.

The Peanut Economy looks at this and celebrates:

"Excellent. Now we can produce more."

But more of what?

More insight?

More truth?

More originality?

More courage?

More wisdom?

Or merely more polished output?

If AI is placed inside a culture that already worships speed, it will not automatically deepen that culture. It will accelerate the shallowness already present.

A shallow organization with AI becomes a faster shallow organization.

A lazy thinker with AI becomes a more productive lazy thinker.

A platform addicted to engagement with AI becomes a larger engine of synthetic noise.

A workplace obsessed with productivity theater becomes a theater with better lighting.

AI does not rescue a culture from its values.

It amplifies them.

That is why Chapter 13 must move beyond AI. The real problem is not the tool alone. The real problem is the economy of attention, incentive, and status into which the tool has been inserted.

AI becomes dangerous when it serves a peanut culture.

THE CHEAPENING OF THOUGHT

Thought becomes cheap when it is separated from cost.

Not financial cost.

Human cost.

The cost of reading.

The cost of attention.

The cost of doubt.

The cost of revision.

The cost of verification.

The cost of silence.

The cost of being wrong.

The cost of beginning again.

The cost of asking a better question.

The cost of saying something original before the world has approved it.

When thought has no cost, it often has no weight.

The Peanut Economy encourages weightless thinking. It allows people to generate opinions without study, content without conviction, strategies without stakes, and language without lived understanding.

This is why everything begins to feel strangely thin.

There is more content than ever, yet less nourishment.

More advice than ever, yet less wisdom.

More commentary than ever, yet less courage.

More information than ever, yet less judgment.

More productivity than ever, yet less meaning.

The world is not suffering from a shortage of output.

It is suffering from a shortage of earned insight.

THE COLLAPSE OF ATTENTION

Attention is the soil of thought.

If attention is constantly disturbed, thought cannot deepen.

The Peanut Economy is built on attention fragmentation. Notifications, feeds, alerts, messages, tabs, videos, shorts, headlines, reactions, replies, and endless updates pull the mind into a thousand shallow contacts with reality.

The result is not just distraction.

It is deformation.

A distracted mind eventually becomes a mind that struggles to remain with anything long enough to understand it. It starts craving novelty. It becomes restless in silence. It treats difficulty as a signal to escape. It feels withdrawal when nothing is happening.

Deep thinking begins to feel unnatural.

That is the tragedy.

The mind was built for depth, but trained for skimming.

It was built for pattern, but trained for interruption.

It was built for contemplation, but trained for reaction.

The Peanut Economy does not merely steal time.

It steals the conditions under which serious thought can form.

And then it sells productivity tools to help you manage the damage.

THE FALSE HERO: THE HYPER-PRODUCTIVE PERSON

Every culture has heroes.

The Peanut Economy worships the hyper-productive person.

The person who wakes at 5 a.m.

CLEARS THE INBOX.

POSTS DAILY.

TRACKS EVERY HABIT.

READS SUMMARIES OF FIVE BOOKS BEFORE BREAKFAST.

RUNS MEETINGS.

SHIPS CONTENT.

OPTIMIZES SLEEP.

AUTOMATES WORKFLOWS.

MAINTAINS A PERSONAL BRAND.

BUILDS A SIDE HUSTLE.

RESPONDS INSTANTLY.

LOOKS UNSTOPPABLE.

BUT THE QUESTION REMAINS:

For what?

Productivity without wisdom is merely accelerated confusion.

A person can be extremely productive and still avoid the one thought that matters. An organization can be highly efficient and still move in the wrong direction. A creator can post daily and never say anything unforgettable. A leader can optimize the system and never ask whether the system deserves to exist.

The Peanut Economy loves the hyper-productive person because they keep the machine fed.
But the paradigm thinker may look less productive from the outside.
They read slowly.
They disappear into research.
They test assumptions.
They reject easy frames.
They spend time in silence.
They produce less frequently but with more force.
They are not idle.
They are gestating.
The Peanut Economy often cannot tell the difference between laziness and incubation.
That is one of its great failures.

THE INDUSTRIALIZATION OF MEDIOCRITY

Mediocrity used to have limits.
A mediocre writer could only produce so much.
A mediocre designer could only make so many weak concepts.
A mediocre strategist could only generate so many shallow plans.
Now mediocrity has machinery.
AI, templates, platforms, automation tools, scheduling systems, analytics engines, and content pipelines can scale average thinking to extraordinary volume.
This is why the future may become flooded with competent emptiness.
Not terrible work.
Not obviously bad work.
Competent emptiness.
Clean essays with no soul.
Professional decks with no insight.
Generated images with no necessity.
Strategies with no courage.
Educational content with no depth.
Brand language with no belief.
Leadership messages with no moral weight.
Everything acceptable.
Nothing unforgettable.

This is the nightmare of the Peanut Economy.
Not a world full of stupidity.
A world full of polished adequacy.
Because polished adequacy is harder to reject than obvious failure. It passes. It gets approved. It fills the calendar. It satisfies the metric. It lowers the standard quietly.
And one day a culture wakes up surrounded by work that is fine.
Only fine.
Terribly, efficiently, beautifully fine.

THE DIFFERENCE BETWEEN OUTPUT AND INSIGHT

Output is what leaves the system.
Insight is what changes the system.
Output can be counted.
Insight must be recognized.
Output fills space.
Insight shifts perspective.
Output answers the question.
Insight changes the question.
Output can be generated quickly.
Insight usually requires collision: between evidence and assumption, between experience and theory, between desire and reality, between what is convenient and what is true.

This is why the Peanut Economy prefers output.
Output is manageable. It can be scheduled, measured, packaged, and sold.
Insight is disruptive. It may force a company to abandon a product. It may force a person to change a belief. It may force a field to admit its foundations are outdated. It may force a culture to confront its contradictions.
Insight is not always convenient.
Therefore, the Peanut Economy underpays it.
But every meaningful human advance depends on it.
A society that rewards output while starving insight becomes noisy, efficient, and directionless.
It produces more maps but forgets where it is going.

THE REBELLION AGAINST THE PEANUT ECONOMY

The answer is not to abandon tools, platforms, AI, or productivity.
The answer is to redesign our relationship with them.
We need an intellectual rebellion.
Not a theatrical rebellion.
Not a fashionable rebellion.
A practical rebellion.
A rebellion of attention, standards, and judgment.
The goal is not to become slow for the sake of slowness. The goal is to restore the conditions under which depth can exist.
Sometimes speed is useful.
Sometimes summaries are useful.
Sometimes automation is useful.
Sometimes platforms are useful.
But none of them should be allowed to define the value of thought.
The sovereign mind uses speed as a servant.
The peanut mind accepts speed as a master.

PRACTICE ONE: INTELLECTUAL MONASTICISM

Every serious thinker needs protected space where the Peanut Economy cannot enter.
Call it intellectual monasticism.
This does not mean withdrawing from the world forever. It means creating deliberate zones of resistance against digital noise.
No feed.
No notifications.
No summaries.
No algorithmic recommendations.
No instant commentary.
No synthetic urgency.
Just the human mind, a difficult source, and enough silence for thought to begin.
This is not nostalgia.
It is cognitive hygiene.
A person who never leaves the stream eventually forgets how to think outside the current.
Try a hard-copy discipline.
Print an essay.
Read a physical book.

Write notes by hand.
Sit with one question for an hour without searching for the answer.
Let the mind ache a little.
That ache is not failure.
It is attention rebuilding muscle.
The Peanut Economy says friction is inefficient.
The sovereign mind says friction is where depth returns.

PRACTICE TWO: THE ANTI-VOLUME MANIFESTO

The Peanut Economy asks:
How much did you produce?
The sovereign mind asks:
What did your work change?
This is the anti-volume manifesto.
Imagine being allowed to send only three emails in a week.
Imagine being allowed to write only one report.
Imagine being allowed to publish only one idea.
Suddenly the question changes.
You cannot hide behind volume.
You must choose.
What is the one sentence that matters?
What is the one insight worth defending?
What is the one contradiction everyone is avoiding?
What is the one design move that changes the whole system?
What is the one question that makes the current strategy look small?
This discipline forces compression.
Not shallow compression.
Explosive compression.
The kind of compression found in a seed, a mantra, a theorem, a great line of poetry, a design principle, a legal argument, or a revolutionary thesis.
The Peanut Economy produces bulk.
The sovereign mind produces force.

PRACTICE THREE: THE ALGORITHMIC HIJACK

Most people allow platforms to build a predictable version of

themselves.
The algorithm learns what you like.
Then it gives you more of it.
Then your world narrows.
Then your taste hardens.
Then your assumptions become invisible.
Then you call the cage "my preference."
The algorithmic hijack is a deliberate attempt to break that cage.
Seek what does not flatter you.
Read outside your profession.
Study a field that makes your own field uncomfortable.
Expose yourself to forms, histories, aesthetics, philosophies, and methods that disturb your default patterns.
If you are a minimalist designer, study maximalist temple architecture.
If you are a technologist, study ritual.
If you are a business strategist, study ecology.
If you are a data person, study poetry.
If you are a writer, study engineering.
If you are a musician, study urban planning.
If you are a leader, study children's play.
This is not random consumption.
It is creative cross-pollination.
Paradigm shifts often emerge when distant fields collide.
The Peanut Economy gives you more of yourself.
The paradigm mind seeks productive strangers.

PRACTICE FOUR: THE SOVEREIGN VERIFICATION RULE

The Peanut Economy encourages people to speak before they know.
The sovereign mind reverses this.
Adopt a rule:
If I cannot explain it without the screen, I do not yet own it.
This does not mean you must be an expert on everything. It means you must stop pretending.
There is dignity in saying:
"I do not know yet."
"I have only read a summary."

"I need to check the source."
"I am not ready to take a position."
"My understanding is incomplete."
These sentences are acts of intellectual hygiene.
They protect you from becoming a distributor of borrowed certainty.
Before expressing a strong opinion, ask:
Can I explain the underlying issue?
Can I describe the strongest opposing view?
Can I name the evidence?
Can I identify what would change my mind?
Can I separate what I know from what I merely heard?
If not, pause.
The Peanut Economy fears the pause.
The sovereign mind uses the pause as a gatekeeper.

PRACTICE FIVE: THE DEEP WORK LEDGER

The Peanut Economy counts visible output.
So create a private counter-system.
Track depth.
Not to become mechanical, but to remind yourself what matters.
Ask at the end of the week:
What did I understand more deeply?
What assumption did I challenge?
What difficult source did I engage?
What did I verify instead of repeat?
What did I refuse to publish because it was not ready?
What idea did I sit with longer than was comfortable?
What did I change my mind about?
What silence did I protect?
This is the ledger the Peanut Economy does not keep.
But it may be the only ledger that saves your mind.

PRACTICE SIX: THE ONE-HARD-QUESTION MEETING

Most meetings are factories of mild agreement.
People gather, perform alignment, avoid tension, and leave with action items that preserve the existing frame.
To resist this, every serious meeting should contain one hard

question.
Not a decorative question.
A dangerous one.
For example:
What are we pretending not to know?
What assumption would destroy this plan if it were false?
What are we optimizing that should not exist?
Who pays the cost of our efficiency?
What would a courageous version of this decision look like?
Which metric is making us stupid?
What would we stop doing if we were honest?
One hard question can puncture productivity theater.
One hard question can save a team from months of polished nonsense.
The Peanut Economy wants meetings to produce tasks.
The sovereign organization wants meetings to produce clarity.

PRACTICE SEVEN: BUILD A PARADIGM STUDIO

Every individual, team, school, studio, or organization needs a protected space for non-peanut thinking.
Call it a Paradigm Studio.
This is where output pressure is temporarily suspended so deeper inquiry can happen.
In the Paradigm Studio, the goal is not to produce more.
The goal is to reframe.
The team asks:
What is the hidden structure?
What are the inherited assumptions?
What contradiction is becoming unbearable?
What does the current system make impossible?
What would we design if we were not afraid of breaking the old model?
What would this look like if dignity, beauty, learning, silence, trust, or courage became the central metric?
The Paradigm Studio is not a brainstorming room full of sticky notes and corporate snacks.
It is a disciplined environment for frame-breaking.
It is where AI can be used properly: not to mass-produce content, but to test assumptions, compare worldviews, generate alterna-

tives, simulate futures, expose contradictions, and challenge lazy answers.
In other words, the Paradigm Studio is the antidote to the Peanut Economy.

THE HUMAN BEING AS MORE THAN A CONTENT NODE

The deepest insult of the Peanut Economy is that it treats the human being as a content node.

CONSUME.
REACT.
PRODUCE.
POST.
MEASURE.
REPEAT.

But a human being is not a node.
A human being is a meaning-maker.
A human being is capable of reverence, doubt, grief, wonder, responsibility, invention, devotion, sacrifice, memory, and moral courage.
The Peanut Economy shrinks this vast interior world into engagement behavior.
It turns the person into a profile.
It turns the mind into a data pattern.
It turns creativity into output.
It turns conviction into branding.
It turns learning into content consumption.
This is why the rebellion must be more than professional.
It must be existential.
To reject the Peanut Economy is to insist that the human mind is not valuable only when it produces visible material.
The mind is also valuable when it contemplates.
When it listens.
When it refuses.
When it waits.
When it connects.
When it grieves.
When it wonders.

When it changes.
When it sees.
A civilization that forgets this will become rich in content and poor in soul.

THE BRIDGE TO ETHICS

This chapter brings us to the edge of a deeper moral question.
If the Peanut Economy rewards output over insight, then people will be tempted to produce without understanding, publish without checking, speak without conviction, and decide without responsibility.
That is why the next question becomes ethical.
Who owns a thought that was generated by a machine?
Who is responsible when automated productivity causes human harm?
What happens when people stop forming their own convictions and merely rent opinions from platforms, models, and feeds?
The Peanut Economy creates the conditions for ethical evasion.
It teaches people to move fast.
Then it tempts them to hide when speed causes damage.
This is why Chapter 14 must follow.
Because once thought becomes cheap, responsibility becomes slippery.
And once responsibility becomes slippery, judgment itself is in danger.

CLOSING: TEAR UP THE PEANUT BALANCE SHEET

The Peanut Economy survives because we keep accepting its exchange rate.
It offers speed in exchange for attention.
Visibility in exchange for depth.
Convenience in exchange for judgment.
Productivity in exchange for meaning.
Confidence in exchange for truth.
Output in exchange for insight.
And too often, we accept.
We accept because the bargain feels efficient.

We accept because the platforms reward us.
We accept because the workplace praises us.
We accept because the feed validates us.
We accept because thinking deeply is lonely, slow, and difficult.
But the balance sheet is fraudulent.
YOU ARE NOT MERELY SAVING TIME.
YOU MAY BE LOSING THE CAPACITY TO USE TIME WELL.
YOU ARE NOT MERELY PRODUCING MORE.
YOU MAY BE WEAKENING YOUR ABILITY TO PRODUCE WHAT MATTERS.
YOU ARE NOT MERELY STAYING VISIBLE.
YOU MAY BE BECOMING DEPENDENT ON VISIBILITY.
YOU ARE NOT MERELY CONSUMING SUMMARIES.
YOU MAY BE LOSING THE STRENGTH TO READ REALITY DIRECTLY.
YOU ARE NOT MERELY USING TOOLS.
YOU MAY BE LETTING TOOLS DEFINE THE SHAPE OF YOUR MIND.
So here is the final challenge of this chapter:
Stop measuring your mind by how much it produces.
MEASURE IT BY WHAT IT CAN PERCEIVE.
MEASURE IT BY WHAT IT CAN QUESTION.
MEASURE IT BY WHAT IT CAN VERIFY.
MEASURE IT BY WHAT IT CAN IMAGINE.
MEASURE IT BY WHAT IT CAN REFUSE.
MEASURE IT BY WHAT IT CAN STAND BEHIND.
The factory whistle is blowing.
The conveyor belt is moving.
The platforms are hungry.
The dashboards are glowing.
The summaries are waiting.
The machine is ready to generate.
But your mind is not a cheap input in someone else's economy.

YOUR JUDGMENT IS NOT A CONTENT FARM.
YOUR ATTENTION IS NOT A DISPOSABLE RAW MATERIAL.
YOUR INSIGHT IS NOT A PEANUT.

Walk out of the factory.
Build the studio.
Protect the silence.
Ask the hard question.
Do the real reading.
Verify before speaking.
Create less noise and more meaning.
Because the future will not be saved by those who produce the most.
It will be shaped by those who can still think deeply enough to know what is worth producing.

DATA MIND, EMPTY BODY
TECH-DYSTOPIA SYMBOLISM
THE CORPORATE VOID

Part V

Toward a Paradigm Shift

Chapter 14 The Ethics of Mental Ownership

Why Your Judgment Cannot Be Outsourced

We have spent much of this book examining the danger of selling the brain for peanuts.

At first, the danger looked practical.

People were using AI to avoid effort. They were outsourcing writing, summarizing, designing, researching, comparing, deciding. The mind became lazy. The user became lethargic. The machine became a convenient excuse. Judgment began to rust quietly in the background.

Then the danger became intellectual.

AI hallucinated. It repeated the status quo. It polished mediocrity. It made shallow thinking look sophisticated. It gave people instant answers before they had earned the right to believe those answers.

Then the danger became creative.

Would AI become a paradigm studio or a peanut factory? Would it help humans test hidden assumptions and design impossible futures, or would it merely mass-produce the average?

Now the danger becomes moral.

This chapter is not about productivity.

It is not about clever prompting.

It is not about whether AI can help you write faster, design faster, code faster, research faster, or publish faster.

This chapter asks a heavier question:

When you outsource thinking, what happens to responsibility?

Because every thought that becomes an action enters the moral world.

A sentence can influence a person.

A recommendation can shape a career.

A design can manipulate behavior.

A hiring filter can exclude a life.

A medical summary can alter treatment.

A financial decision can damage a family.

A political message can inflame a society.
A classroom tool can shape a child's mind.

So we must stop pretending that AI assistance is merely a technical convenience. It is also an ethical arrangement. The moment a human uses a machine-generated idea in the world, that human becomes part of a chain of responsibility.

And here is the central rule:

You can outsource a task, but you cannot outsource responsibility.

That is the hard moral line.

You may ask AI to draft.
You may ask AI to summarize.
You may ask AI to compare.
You may ask AI to simulate.
You may ask AI to provoke.

But if you act on the result, publish the result, teach the result, sell the result, approve the result, recommend the result, or impose the result on another human being, then the burden returns to you.

The machine may generate the text.

You own the consequence.

THE NEW MORAL CRISIS

The old world had a simple moral structure.

A person thought something, said something, wrote something, designed something, decided something, and then could be held accountable for it.

The chain was not always perfect, but it was visible.

Now the chain has become foggy.

A human gives a prompt. A machine generates an answer. The human copies it. Another system modifies it. A platform distributes it. A user believes it. A decision follows. A person is helped, harmed, persuaded, rejected, misled, diagnosed, approved, denied, hired, fired, or forgotten.

And when harm appears, everyone points elsewhere.

The user says, "The AI gave it to me."
The company says, "The model produced it."
The developer says, "The user misapplied it."
The institution says, "The workflow approved it."
The manager says, "The system recommended it."

The machine says nothing.
Because the machine has no conscience.
This is the ethical vacuum of outsourced judgment.
The more complex the chain becomes, the easier it is for humans to hide inside it. Responsibility dissolves into process. Moral agency gets buried under dashboards, terms of service, automation logs, and confidence scores.
But the person harmed by the decision does not suffer statistically. They suffer personally.
The rejected applicant does not care that the model was "only a tool."
The misdiagnosed patient does not care that the summary was "AI-assisted."
The student taught a false idea does not care that the teacher "trusted the generated content."
The customer manipulated by automated persuasion does not care that the interface was "optimized."
Harm lands on flesh, not on software.
That is why judgment cannot be outsourced.

MENTAL OWNERSHIP: THE MISSING CONCEPT

We often speak about intellectual property.
Who owns the copyright?
Who owns the output?
Who owns the data?
Who owns the model?
Who owns the platform?
These questions matter, but they are not enough.
There is a deeper idea we must recover:
mental ownership.
Mental ownership is not merely legal ownership. It is the inner relationship between a human being and a thought.
A person owns a thought morally when they have examined it, tested it, understood it, accepted its implications, and are willing to stand behind it.
A thought is not truly yours simply because you possess the sentence.

A thought is yours when you have wrestled with it.
A machine can hand you a paragraph.
But it cannot hand you conviction.
A machine can generate a position.
But it cannot give you moral authorship.
A machine can arrange arguments.
But it cannot make you answerable before truth.
This is where the peanut mind collapses. It thinks possession is ownership.
It says:
"I have the output, therefore I have the idea."
No.
You may have the output.
But if you cannot explain it, defend it, verify it, revise it, and accept responsibility for its consequences, then you do not own the thought.
You are merely carrying borrowed language.
And borrowed language becomes dangerous when it walks around wearing the costume of conviction.

THE PHANTOM AUTHOR

Who owns a thought born from code?
This question is more complicated than it first appears.
If a person types a lazy prompt and AI produces a beautiful essay, is the person the author?
Technically, perhaps in some contexts.
Morally, not yet.
Authorship is not only the act of causing words to appear. Authorship is the act of giving form to meaning through judgment.
If you ask AI to generate a speech about justice and then read it without understanding the arguments, without believing the claims, without checking the facts, without feeling the moral weight of the words, you may be the speaker, but you are not fully the author.
You are the mouthpiece of a borrowed arrangement.
This is the rise of the phantom author.
The phantom author signs the work but did not form the thought.

The phantom author publishes the article but cannot defend the argument.
The phantom author submits the essay but has not internalized the learning.
The phantom author presents the strategy but does not understand the assumptions.
The phantom author declares a belief but has never suffered the burden of choosing it.
This is not merely dishonest.
It is spiritually hollow.
Because the point of thinking is not just to produce statements.
The point of thinking is to become the kind of person who can stand inside those statements with integrity.
When people stop forming their own thoughts, they do not merely lose originality.
They lose interiority.
They become externally fluent and internally vacant.

THE DIFFERENCE BETWEEN ASSISTANCE AND ABDICATION

We must be careful here.
This chapter is not saying that using AI makes a person morally fake.
That would be too simple and too foolish.
Human thought has always been assisted.
We think with books.
We think with teachers.
We think with conversations.
We think with libraries.
We think with instruments.
We think with diagrams, calculators, notebooks, search engines, mentors, critics, and communities.
No serious thinker is completely self-made.
The question is not whether thought is assisted.
The question is whether thought is abdicated.
Assistance strengthens agency.
Abdication replaces agency.
Assistance says:

"Help me examine this."
Abdication says:
"Tell me what to think."
Assistance says:
"Give me material I can judge."
Abdication says:
"Give me an answer I can avoid judging."
Assistance says:
"Challenge my assumptions."
Abdication says:
"Relieve me from the discomfort of having assumptions."
The Stewardship Mind welcomes assistance.
But it rejects abdication.
The ethical line is crossed when the human stops participating in the formation of belief and simply accepts the convenience of generated certainty.

THE SHIELD OF THE CLERK

One of the oldest tricks of human cowardice is to hide behind an instrument.
"I was only following the rule."
"I was only using the system."
"I was only applying the policy."
"I was only trusting the data."
In the AI age, this becomes more seductive:
"The model recommended it."
This is the Shield of the Clerk.
It is the attempt to convert moral responsibility into administrative obedience.
Imagine a hiring manager using an AI tool to screen thousands of applicants. The tool ranks candidates based on patterns from past company data. It appears efficient. It appears objective. It appears neutral.
But the historical data carries old prejudice. The model quietly downgrades candidates from certain neighborhoods, career gaps, accents, institutions, or nontraditional backgrounds. Qualified people disappear before a human ever reads their names.

When challenged, the manager says:
"I did not discriminate. The system filtered them."
This is not innocence.
This is outsourced prejudice with a software interface.
The model may have processed the data, but the institution chose to use it. The manager chose to trust it. The company chose the workflow. The human decision-makers chose not to inspect the consequences.
A machine can automate a bias.
Only humans can be accountable for allowing that bias to become policy.
The same principle applies everywhere.
A teacher who uses AI-generated material still owns what is taught.
A doctor who uses AI-assisted summaries still owns clinical judgment.
A lawyer who uses AI-generated citations still owns legal accuracy.
A designer who uses AI-generated patterns still owns the user impact.
A leader who uses AI-generated strategy still owns the organizational consequence.
A citizen who spreads AI-generated misinformation still owns the harm of spreading it.
The machine can be part of the process.
It cannot become the moral scapegoat.

THE AUTOMATED JUDGE

Consider the tragedy of Judge Barnaby.
Barnaby is a magistrate in a busy municipal court. His docket is overflowing. The system is underfunded. The public is impatient. Every day, he faces more cases than any human being can carefully absorb.
Then the court introduces an AI-powered legal assistant called JusticeBot.
JusticeBot is fast, elegant, and confident. It summarizes case histories, compares sentencing patterns, calculates risk, cites precedent, and produces recommendations in polished legal language.

At first, Barnaby feels relieved.
The backlog shrinks.
The reports look professional.
The court appears modern.
Then one morning, a twenty-two-year-old man appears before him. The young man has been caught stealing bread and basic groceries. It is not a glamorous case. It is not a dramatic case. It is the kind of case bureaucracies process quickly because poverty rarely has good public relations.
JusticeBot reviews the file.
It analyzes neighborhood statistics, prior policing data, historical sentencing patterns, employment instability, school records, and demographic correlations. In less than a second, it produces a recommendation:
High risk of reoffending.
Recommend severe custodial sentence.
Three years.
The recommendation is beautifully formatted.
The legal reasoning sounds coherent.
The risk chart looks scientific.
Barnaby is tired.
He does not ask whether the policing data was biased. He does not ask whether poverty is being converted into criminal risk. He does not ask whether historical injustice is being laundered through statistical language. He does not look carefully at the young man's circumstances. He does not ask what justice requires.
He sees fluency.
He sees confidence.
He sees efficiency.
Then he lowers the gavel.
Three years.
Later, an investigation reveals that JusticeBot was trained on decades of biased sentencing patterns. The system did not discover justice. It reproduced historical punishment with mathematical polish.
The young man's life has been damaged.
The court is embarrassed.
Barnaby defends himself:

"I followed the system recommendation."
But that is precisely the failure.
A judge is not paid to decorate machine output with a robe.
A judge exists to judge.
The tragedy is not that Barnaby used AI.
The tragedy is that he let AI occupy the moral space where his conscience should have stood.
He outsourced analysis and then accidentally outsourced humanity.

THE DEATH OF CONVICTION

There is another danger, quieter but more profound.
What happens when people stop forming their own convictions?
A conviction is not an opinion you downloaded.
It is not a slogan you borrowed.
It is not a paragraph generated in your preferred tone.
A conviction is a belief that has passed through friction.
It has been tested by doubt.
It has survived contradiction.
It has cost you something.
It has become part of your moral spine.
Real convictions are not instantly produced. They are forged through reading, suffering, listening, failure, reflection, argument, repentance, and courage.
AI can help examine convictions.
But it cannot forge them for you.
If people begin outsourcing their moral positions to machines, they may still sound articulate. They may still post statements, write essays, join debates, and present polished arguments. But inside, the belief may be hollow.
This creates a terrifying new kind of person:
The person with fluent opinions and no formed conscience.
Such a person is easy to manipulate.
Change the feed, and their beliefs shift.
Change the model output, and their confidence shifts.
Change the dominant language, and their morality shifts.
They become rhetorically strong but spiritually weak.

The danger is not that they have no opinions.
The danger is that they have too many opinions and no ownership of any of them.
A civilization full of such people becomes unstable. It can be moved by the next persuasive system, the next optimized narrative, the next synthetic consensus, the next beautifully generated lie.
Conviction is not optional.
It is the immune system of the moral mind.

BORROWED BELIEFS AND SYNTHETIC CERTAINTY

AI can create synthetic certainty.
It can take a complex issue and produce a clean answer with strong language, tidy structure, and persuasive rhythm. The user feels clarity. The anxiety drops. The discomfort fades.
But sometimes that clarity is not earned.
It is manufactured.
This matters because moral questions are often difficult for a reason. They involve competing goods, conflicting duties, historical wounds, personal stakes, and unknown consequences. If an answer arrives too cleanly, it may have removed the very tension that made the question ethical.
A serious moral mind must resist premature certainty.
It must ask:
What pain is hidden behind this policy?
Who is absent from this answer?
Who benefits if I accept this framing?
Who pays the cost if I am wrong?
What does this answer make invisible?
What would I believe if I had to face the harmed person directly?
AI can help generate these questions.
But the human must feel their weight.
Without that weight, ethics becomes styling.
A moral paragraph without moral ownership is just decorative virtue.

THE RESPONSIBILITY CHAIN

To preserve mental ownership, we need a clear responsibility chain.

Before using AI-assisted output in any meaningful context, ask five questions:

1. Who prompted the system?

The prompt matters because it frames the result.

A biased question can produce a biased answer.

A lazy question can produce shallow output.

A manipulative question can produce harmful persuasion.

The person who frames the question already carries responsibility.

2. Who selected the output?

AI may generate options, but the human chooses what to use.

Selection is judgment.

If you choose the most convenient output because it flatters your goal, you own that choice.

3. Who verified the claim?

If the output contains facts, citations, numbers, recommendations, risks, or technical claims, verification is not optional.

Failure to check is not a harmless oversight.

In high-stakes contexts, failure to check can become negligence.

4. Who applied the result?

The moment output becomes action, responsibility intensifies.

A draft is one thing.

A decision is another.

A suggestion is one thing.

A policy is another.

An experiment is one thing.

A deployment is another.

5. Who bears the consequence?

This is the final moral test.

If the people affected cannot hold the machine accountable, then they must be able to hold a human accountable.

Where there is consequence, there must be answerability.

No ethical system should allow responsibility to disappear into automation.

The answer is not to reject AI.
That would be too small.
The answer is to use AI under ethical discipline.
Here are the duties of the mentally sovereign user.

The Duty to Disclose When It Matters

Not every AI-assisted sentence requires a confession.
But in contexts where trust, authorship, expertise, evaluation, or public consequence matters, disclosure becomes important.
A student should not pretend machine-generated work is personal learning.
A researcher should not hide AI-generated analysis that shaped conclusions.
A professional should not conceal automation where human review is expected.
A company should not pretend a decision was fully human if an AI system materially influenced it.
Disclosure protects trust.
It also protects the user from becoming a phantom author.

The Duty to Verify

Verification is the price of using a powerful tool.
If AI helps produce a claim, the human must check the claim.
The more important the claim, the stronger the verification required.
This is especially true for law, medicine, finance, safety, education, public policy, history, and science.
The machine may assist.
The human must confirm.

The Duty to Understand

Do not use what you cannot explain.
This is a brutal but necessary rule.
If you cannot explain the reasoning behind a recommendation, you should be cautious about acting on it. If you cannot explain the assumptions behind an argument, you should be cautious about publishing it. If you cannot explain the logic of a decision, you should be cautious about imposing it on others.
Understanding is not decoration.
It is part of ethical ownership.

The Duty to Preserve Human Appeal

Any AI-assisted decision that affects a person's life should allow meaningful human review.

People should not be trapped inside automated conclusions.

There must be a door back to human judgment.

A rejected applicant should be able to appeal.

A student flagged by a system should be heard.

A patient affected by an algorithmic summary should receive human attention.

A citizen scored by a system should have recourse.

A civilization that removes appeal removes dignity.

The Duty to Protect the Vulnerable

AI systems often magnify existing power.

Those with money, data, infrastructure, and institutional authority can automate decisions at scale. Those affected by those decisions may have little visibility or control.

Therefore, ethical AI use must ask:

Who is most vulnerable if this system fails?

Who has no voice in this design?

Who will be reduced to a pattern?

Who will be misclassified?

Who will be unable to contest the result?

The vulnerable should not become test subjects for the convenience of the powerful.

The Duty to Keep Forming Convictions

Perhaps the deepest duty is this:

Continue becoming a person.

Read.

Reflect.

Argue.

Listen.

Revise.

Admit error.

Hold tension.

Choose consciously.

Do not let the machine become the womb of your worldview.

Use it as a tool, but do not let it become the parent of your convictions.

THE MENTAL OWNERSHIP TEST

Before accepting an AI-assisted idea as your own, apply this test.
Ask yourself:

CAN I EXPLAIN THIS IN MY OWN WORDS?
CAN I IDENTIFY THE ASSUMPTIONS BEHIND IT?
CAN I VERIFY THE IMPORTANT CLAIMS?
CAN I NAME WHO MIGHT BE HARMED IF THIS IS WRONG?
CAN I DEFEND THIS WITHOUT HIDING BEHIND THE MACHINE?
CAN I REVISE IT FROM MY OWN UNDERSTANDING?
CAN I REJECT IT IF MY CONSCIENCE DISAGREES?
CAN I STAND BEFORE ANOTHER HUMAN BEING AND OWN THE CONSEQUENCE?

If the answer is no, the thought is not fully yours yet.
It may be useful.
It may be promising.
It may be beautifully written.
But it has not crossed the threshold into moral ownership.
Do not sign your name to it too quickly.

THE DIFFERENCE BETWEEN OUTPUT AND INTEGRITY

The Peanut Economy will reward output.
It will reward those who can generate more proposals, more posts, more designs, more emails, more reports, more strategies, more images, more slogans, more noise.
But integrity is not measured by output.
Integrity is measured by the relationship between what you produce and what you actually understand, believe, verify, and stand behind.
A world drowning in generated output will become hungry for trustworthy humans.
Not merely productive humans.
Trustworthy humans.
Humans who can say:
"I checked this."
"I understand this."
"I disagree with the machine here."
"I will not publish that claim."

"I will not automate that decision."
"I will not hide behind the system."
"I am responsible."
That sentence—"I am responsible"—may become one of the most radical sentences of the AI age.

THE MORAL FAILURE OF "GOOD ENOUGH"

In earlier chapters, we explored how "good enough" becomes a disease.
Here, that disease becomes ethical.
In low-stakes work, "good enough" may create boredom.
In high-stakes work, "good enough" can create harm.
A good-enough summary can mislead.
A good-enough policy can exclude.
A good-enough diagnosis can miss a life-threatening detail.
A good-enough risk score can punish the wrong person.
A good-enough educational tool can misform a child's understanding.
A good-enough public message can distort reality.
The peanut mind says:
"It is probably fine."
The ethical mind says:
"Probably is not enough when someone else pays the price."
This is the moral upgrade required now.
We must learn to distinguish between tasks where speed is acceptable and decisions where care is mandatory.
Not every use of AI requires solemn ceremony.
But some uses do.
The mature mind knows the difference.

THE SACRED BOUNDARY

There are areas of life where AI can assist but must never rule.
JUSTICE.
CARE.
EDUCATION.
GRIEF.
LOVE.
FAITH.
DEMOCRACY.

HUMAN DIGNITY.
MORAL FORMATION.
CREATIVE MEANING.
In these areas, efficiency must bow before humanity.
A machine can help organize evidence in a legal case, but it must not replace the moral act of judgment.
A machine can help summarize medical records, but it must not replace the human duty of care.
A machine can help prepare lessons, but it must not replace the teacher's responsibility to awaken minds.
A machine can help write words of sympathy, but it must not replace the actual act of compassion.
A machine can help explore ethical arguments, but it must not become the conscience.
The boundary is simple:
Use AI to support human presence, not substitute for it.
When AI removes drudgery so humans can become more attentive, it serves dignity.
When AI removes humans from places where presence is morally required, it attacks dignity.

MENTAL OWNERSHIP IN CREATIVE WORK

For writers, artists, designers, musicians, and thinkers, the ethics of mental ownership has a special intensity.
Creative work is not just output. It is identity shaped into form.
This does not mean AI cannot assist creativity. It can. It can generate variations, suggest structures, challenge clichés, expand references, and provoke new directions.
But the creator must remain awake.
The danger is not that AI will help create.
The danger is that creators will stop knowing why they chose what they chose.
A designer who cannot explain the design is not yet the author of the design.
A writer who cannot defend the sentence is not yet the author of the sentence.
A musician who cannot feel the necessity of the phrase is not yet

the owner of the phrase.
A strategist who cannot explain the tradeoff is not yet the owner of the strategy.
Creative ownership requires taste.
Taste is judgment under pressure.
It is the ability to say yes, no, less, more, this, not that, too easy, too false, too polished, too dead, this one has life.
AI can generate options.
Taste must choose.
That choosing is where authorship lives.

MENTAL OWNERSHIP IN LEADERSHIP

Leadership may be the area where outsourced judgment becomes most dangerous.
A leader's decisions affect other people's time, money, dignity, safety, and future. If leaders begin using AI to generate strategy without understanding the human consequences, organizations will become more efficient and less wise.
A leader cannot say:
"The model told us to cut the team."
"The system recommended the policy."
"The dashboard optimized the decision."
Leadership is not the art of obeying dashboards.
Leadership is the burden of deciding what the dashboard cannot know.
The dashboard may show cost.

It may not show humiliation.
The model may show productivity.
It may not show burnout.
The system may show risk.
It may not show courage.
The report may show efficiency.
It may not show trust.
This is why leadership requires mental ownership. A leader must

understand not only what is recommended, but what is being sacrificed.
AI can inform leadership.
It cannot become leadership.

THE CHAPTER MANTRA

The mantra of this chapter is:
If you do not own the process of your thinking,
you have no right to claim ownership over the path of your life.
This is not merely a warning.
It is a call to adulthood.
Because a person who does not own their thinking becomes easy to govern from the outside. Their beliefs can be shaped by feeds. Their desires can be shaped by recommendation engines. Their language can be shaped by generated templates. Their moral instincts can be replaced by whatever answer arrives most fluently.
Mental ownership is freedom.
Without it, the mind becomes rented property.
And the landlord may not have your soul in mind.

Closing: Responsibility Is the Last Human Crown
The age of AI will test many things.

It will test education.
It will test business.
It will test art.
It will test law.
It will test politics.
It will test the meaning of expertise.

But above all, it will test responsibility.

When machines become powerful enough to assist nearly every act of thought, humans will face a temptation older than technology itself:
The temptation to escape blame.
To say:
"It was not me."
"It was the system."
"It was the model."
"It was the recommendation."
"It was the process."

But the sovereign human must refuse that escape.
The ethical mind says:
"I used the tool.
I selected the output.
I checked or failed to check.
I acted.
I am answerable."
This is not a burden to resent.
It is the final dignity of being human.

A machine can produce.
A human must answer.
A machine can calculate.
A human must care.
A machine can recommend.
A human must decide.
A machine can simulate thought.
A human must own judgment.

So do not sell your brain for peanuts.
Do not rent out your conscience to a statistical engine.
Do not become the phantom author of your own life.
USE THE MACHINE.
QUESTION THE MACHINE.
CHALLENGE THE MACHINE.
LEARN FROM THE MACHINE.

But when the moment of consequence arrives, stand where only a human can stand.
Stand inside responsibility.
Because you can outsource a task.
But you cannot outsource the moral ownership of your mind.

Chapter 15
The Human Upgrade

A Manifesto for Thinking in the Age of AI

We have now travelled through the full anatomy of cognitive surrender.

We began with the peanut bargain: the quiet temptation to exchange the hard work of thinking for the cheap comfort of instant output. We watched the flimsy mind avoid doubt, discomfort, contradiction, and depth. We saw how "good enough" becomes a mental disease when people stop asking whether an answer is true, meaningful, original, or responsible.

We entered the hallucination fog, where fluent nonsense wears the costume of knowledge. We examined the lethargic user, the outsourced mind, the status quo machine, and the danger of allowing AI to repeat the past instead of helping us reimagine the future.

Then we turned toward recovery.

We developed the verification ritual. We transformed prompting into provocation. We reclaimed stewardship. We asked whether AI would become a paradigm studio or a peanut factory. We exposed the Peanut Economy, where platforms, speed, productivity theater, viral summaries, and shallow expertise reward output instead of insight. Finally, we confronted the ethics of mental ownership and declared the central moral law of this age:

You can outsource a task, but you cannot outsource responsibility.

Now we arrive at the final main chapter.

This chapter is not a warning.

It is a call.

This book was never written to make people fear AI. Fear is too small. Fear makes the human shrink, and the whole purpose of this journey has been to make the human mind stand taller.

AI is not the end of human intelligence.

It is the test of human intelligence.

It is the great filter that will reveal who has merely been carrying information and who has learned how to think. It will reveal who depends on borrowed fluency and who owns judgment. It will reveal who asks for answers and who knows how to interrogate answers until a deeper question appears.

The arrival of AI does not make the human mind obsolete.

It makes a weak mind obvious.

It makes a lazy mind dangerous.

It makes a sovereign mind priceless.

This is the Human Upgrade.

THE HUMAN UPGRADE IS NOT A SOFTWARE UPDATE

The Human Upgrade is not about installing another app, learning another productivity trick, or mastering a fashionable prompt formula.

It is not about becoming faster than the machine. That battle is already lost. You will not out-summarize the model. You will not out-format the model. You will not out-generate the model. You will not out-produce an engine designed for endless production.

The upgrade is not speed.

The upgrade is depth.

The upgrade is not output.

The upgrade is judgment.

The upgrade is not having more answers.

The upgrade is knowing what to do with answers.

A human being who competes with AI on volume becomes a exhausted clerk in the Peanut Factory. But a human being who uses AI to ask sharper questions, test assumptions, expose contradictions, and design impossible futures becomes something else.

A steward.

A critic.

A creator.

A teacher.

A designer of meaning.

The Human Upgrade begins when you stop asking:

How can I get the answer faster?

And begin asking:

What kind of thinker must I become in order to judge this answer well?

That question changes everything.

THE FINAL LESSON OF THE BOOK

Every chapter of this book has been circling one truth:
The mind is not protected by intelligence alone.
Many intelligent people can still become lazy.
Many educated people can still outsource judgment.
Many talented people can still hide behind tools.
Many productive people can still produce nothing of consequence.
The mind is protected by discipline.
By friction.
By verification.
By moral ownership.
By the willingness to say, "I do not know yet."
By the courage to ask, "What assumption am I protecting?"
By the humility to check.
By the imagination to reframe.
By the responsibility to own the result.
The Peanut Contract offered a false promise:
"Give up the burden of thought, and I will give you comfort."
The Human Upgrade offers a harder promise:
"Accept the burden of thought, and you will recover your sovereignty."
There is no shortcut to this sovereignty.
You do not get it by using AI.
You get it by refusing to disappear inside AI.

THE NEW HUMAN ADVANTAGE

In the age of AI, the old human advantage is changing.
Memorization is no longer enough.
Information access is no longer rare.
Basic explanation is no longer special.
Polished writing is no longer proof of deep thought.
Average design is no longer difficult to generate.
Ordinary analysis is no longer a scarce commodity.
So what remains?
What remains is the human ability to judge what should matter.
To decide what is worth doing.
To recognize when a fluent answer is empty.

To sense when a beautiful solution is morally wrong.
To notice when the obvious path is merely the inherited path.
To ask what has been excluded.
To feel the human consequence behind the statistical recommendation.
To create from lived experience, not merely pattern repetition.
To stand inside responsibility.
The new human advantage is not information.
It is discernment.
Not speed.
Discernment.
Not output.
Discernment.
Not fluency.
Discernment.
Discernment is the ability to separate signal from noise, truth from polish, meaning from performance, courage from conformity, and insight from content.
This is the skill the Peanut Economy cannot mass-produce.
This is the skill AI can assist but never own.

THE SOVEREIGN MIND

The goal of this book is not merely the intelligent mind.
It is the sovereign mind.
The intelligent mind can solve problems.
The sovereign mind asks whether the problem has been framed correctly.
The intelligent mind can consume information.
The sovereign mind knows what deserves attention.
The intelligent mind can use AI.
The sovereign mind governs AI.
The intelligent mind can generate options.
The sovereign mind chooses with responsibility.
The intelligent mind may be impressive.
The sovereign mind is answerable.
A sovereign mind does not reject tools. It rejects servitude.
It says:
I will use the machine, but I will not become its extension.

I will accept assistance, but I will not abdicate authorship.
I will welcome speed, but I will not worship speed.
I will generate options, but I will not confuse options with imagination.
I will ask for answers, but I will question them before I believe them.
I will use AI to strengthen my mind, not to outsource it.
This is the posture of the upgraded human.

THE MANIFESTO FOR STUDENTS

Learn the Grappling, Not the Answer

To the student, the temptation is obvious.

AI CAN WRITE THE ESSAY.
AI CAN SUMMARIZE THE BOOK.
AI CAN SOLVE THE PROBLEM.
AI CAN GENERATE THE REFLECTION.
AI CAN PRODUCE THE PROJECT OUTLINE.
AI CAN MAKE YOU LOOK PREPARED.

But education is not the production of assignments.
Education is the formation of the mind.
The assignment is not the point. The struggle is the point.
The wrestling is the point.
The confusion is the point.
The first bad draft is the point.
The moment when you realize your argument is weak is the point.
The painful revision is the point.
The slow reading is the point.
The effort to explain something in your own words is the point.
If AI removes all that, it does not help you learn. It helps you escape learning while preserving the appearance of learning.
That is one of the most dangerous bargains a student can make.
You may graduate with grades, but without grip.
You may collect credentials, but not capacity.
You may know how to submit, but not how to think.
The Human Upgrade for students is simple:
Use AI as a gym, not a cheat sheet.
Write your own argument first. Then ask AI to attack it.
Read the source first. Then ask AI to help you compare interpretations.
Try solving the problem first. Then ask AI where your reasoning

broke.
Draft your own paragraph first. Then ask AI what is unclear.
The goal is not to avoid difficulty.
The goal is to make difficulty more productive.
A student who uses AI to escape the mental workout becomes weaker.
A student who uses AI to intensify the mental workout becomes dangerous in the best way.
The future does not need students who can submit perfect assignments.
It needs people who can think under pressure.

THE MANIFESTO FOR DESIGNERS AND CREATORS

Hunt the Anomaly
To the designer, artist, writer, musician, architect, filmmaker, strategist, or creator, AI offers a glittering temptation:
Instant variation.
Instant moodboards.
Instant drafts.
Instant concepts.
Instant style.
Instant polish.
This can be useful.
It can also be deadly.
Because the easiest thing for AI to generate is the already expected. The probable. The fashionable. The averaged memory of what has already been made. It can produce endless variations of what the world has already taught it to recognize.
But creativity is not the worship of the probable.
Creativity begins when the expected answer becomes unbearable.
The creator's task is not to ask the machine for "a good idea."
The creator's task is to find the strange fracture where a new idea might enter.
Ask AI:
What are the clichés in this field?
What would the average creator do here?
What assumptions does this style protect?
What is the most overused metaphor?

What would happen if I reversed the emotional logic?
What would this become if I combined it with a world that has nothing to do with it?
What is the illogical anomaly that might contain life?
The creator must learn to distrust the first beautiful output.
Beauty can be counterfeit.
Polish can be premature.
Style can become anesthesia.
The Human Upgrade for creators is to use AI not as a replacement imagination, but as a provocation engine.
Let it produce the ordinary so you can reject it faster.
Let it map the consensus so you can escape it.
Let it generate variations so your taste can become sharper.
Let it expose the obvious so you can hunt the anomaly.
Your job is not to produce more.
Your job is to produce what could not have existed without your human collision of memory, pain, intuition, discipline, culture, taste, and risk.
The machine can remix.
You must reveal.

THE MANIFESTO FOR LEADERS

Reject the Shield of the Clerk
To the leader, AI offers efficiency, scale, and decision support.
It can analyze markets.
Summarize reports.
Compare policies.
Draft strategy.
Screen applicants.
Monitor productivity.
Predict risk.
Optimize operations.
But leadership is not the act of obeying recommendations.
Leadership is the burden of judgment when the recommendation is incomplete.
A dashboard may show cost.
It may not show dignity.
A model may show efficiency.

It may not show trust.
An analysis may show probability.
It may not show courage.
A system may show performance.
It may not show meaning.
The great danger for leaders is the Shield of the Clerk: the temptation to hide behind the system when consequences become painful.
"The model recommended the layoffs."
"The algorithm ranked the candidates."
"The system flagged the risk."
"The dashboard showed declining value."
"The AI produced the strategy."
This is not leadership.
This is bureaucratic cowardice with better software.
The Human Upgrade for leaders is stewardship.
Use AI to widen perception, not narrow responsibility.
Use it to test decisions, not excuse decisions.
Use it to reveal blind spots, not bury conscience.
Use it to ask who may be harmed, who may be excluded, who may be misclassified, who may be unheard.
A leader must never allow AI to become the moral laundering machine of the organization.
When the decision affects human lives, the human leader must remain visible.
Visible to the evidence.
Visible to the people affected.
Visible to the possibility of being wrong.
Visible to responsibility.
The upgraded leader says:
I USED THE TOOL.
I EXAMINED THE OUTPUT.
I CONSIDERED THE CONSEQUENCES.
I MADE THE DECISION.
I AM ACCOUNTABLE.
That sentence is leadership in the age of AI.

THE MANIFESTO FOR TEACHERS

Grade the Friction, Not the Polish

To the teacher, AI creates a crisis and an opportunity.

The crisis is obvious: polished work is no longer reliable proof of learning.

A student can submit a clean essay without having wrestled with the topic. A project can look sophisticated without reflecting genuine understanding. A reflection can sound mature while being assembled by a machine.

If teachers continue to evaluate only the final polish, they will reward simulation.

The Human Upgrade for teachers is to grade the process of thought.

Grade the friction.
Grade the questions.
Grade the drafts.
Grade the revisions.
Grade the student's ability to explain why they changed their mind.
Grade source checking.
Grade argument repair.
Grade the ability to identify what AI got wrong.
Grade the courage to say, "My first answer was weak."

Education must move from product inspection to mind formation.

The teacher of the future is not merely a content deliverer. AI can deliver content. The teacher becomes a trainer of judgment.

The classroom must become a place where students learn how to challenge generated answers, detect false fluency, compare perspectives, verify claims, and form convictions.

Do not ask only:

What did the student submit?

Ask:

What did the student struggle with?
What did the student verify?
What did the student reject?
What did the student finally understand that they did not understand before?

The best teachers will not ban tools out of panic or worship tools

out of fashion.
They will build friction architecture.
They will teach students how to remain human while using machines.

THE MANIFESTO FOR ORDINARY PEOPLE

Protect the Everyday Mind
This book is not only for students, designers, leaders, teachers, or professionals.
It is for ordinary people living ordinary days in a world that constantly tries to rent their attention.
The Peanut Economy does not only attack great decisions. It attacks daily perception.
It shapes what you read.
What you skip.
What you believe.
What you desire.
What you fear.
What you repeat.
What you think everyone else thinks.
The ordinary mind is the battlefield.
Every day, you are invited to outsource small acts of judgment.
Let the feed decide what matters.
Let the platform decide what is credible.
Let the summary replace the book.
Let the comments replace reflection.
Let the generated answer replace your own attempt.
Let convenience become your compass.
The Human Upgrade begins in small refusals.
Read something difficult.
Pause before sharing.
Ask, "How do I know this?"
Let silence return.
Write in your own words.
Choose one source over ten fragments.
Admit when you have not thought enough.
Refuse the addictive pleasure of instant certainty.
Your mind is not ordinary because it lacks fame.

Your mind becomes extraordinary when it remains awake in a culture designed to put it to sleep.

THE TEN PRINCIPLES OF THE HUMAN UPGRADE

The Human Upgrade needs principles.

Not vague inspiration.

Operating principles.

1. Use AI as a tool, not a throne.

The machine may assist, but it must not rule. Keep the hierarchy clear. You are the steward. AI is the instrument.

2. Never accept fluency as proof of truth.

A beautiful sentence can still lie. A confident answer can still be wrong. Polish is not evidence.

3. Treat every answer as a beginning.

The first answer should open inquiry, not close it. Ask what assumption shaped it, what evidence supports it, and what possibility it ignores.

4. Preserve friction.

Do not remove all difficulty from learning, creating, deciding, or understanding. Some friction is not inefficiency. Some friction is formation.

5. Verify what matters.

Names, numbers, dates, sources, claims, risks, and high-stakes recommendations must be checked. Verification is respect for reality.

6. Protect your attention.

A fragmented mind cannot think deeply. Guard silence. Guard reading. Guard long thought. Guard the space where insight forms.

7. Form your own convictions.

Do not rent beliefs from feeds, platforms, influencers, or machines. Let your convictions pass through doubt, evidence, experience, and moral reflection.

8. Question the frame.

Do not merely improve the existing answer. Ask whether the question itself is too small.

9. Own the consequence.

If you use the output, you own the decision. Do not hide behind the model.

10. Create meaning, not noise.

The world does not need more content. It needs more insight, courage, beauty, clarity, compassion, and responsible imagination.

These principles are not anti-AI.

They are pro-human.

THE WHOLE-BRAIN AWAKENING

The Human Upgrade is also a whole-brain awakening.

The age of AI will reward neither purely mechanical thinking nor vague intuition alone. It will require integration.

LOGIC AND IMAGINATION.

VERIFICATION AND VISION.

ANALYSIS AND EMPATHY.

STRUCTURE AND MYSTERY.

EVIDENCE AND COURAGE.

Left-brain discipline and right-brain lightning.

The machine is powerful in the world of patterns. It can process vast amounts of prior human expression. It can identify statistical regularities. It can generate probable continuations.

But paradigm shifts often begin where probability fails.

A leap.

A contradiction.

A dream.

A refusal.

A strange analogy.

A moral discomfort.

A pattern that does not fit.

A question nobody wants to ask.

This is where the human whole brain matters.

Not because humans are always rational.

Not because intuition is always correct.

But because human intelligence is embodied, emotional, moral, historical, sensory, relational, and imaginative. We do not merely process data. We suffer consequences. We remember. We grieve. We hope. We love. We build worlds because something inside us cannot accept the present arrangement as final.

The machine can predict the past.

The upgraded human can imagine the necessary.

THE FINAL FLIGHT CHECKLIST

Every time you open an AI tool, pause.

Before the prompt, ask:

Am I escaping effort or sharpening effort?

There is a difference between using AI to remove drudgery and using it to avoid growth.

Am I asking for closure too early?

If the answer arrives before the question has matured, be suspicious.

What assumptions am I feeding the machine?

A lazy prompt often contains a lazy worldview.

What should I verify?

Mark the claims that matter. Do not let polish pass as proof.

What would a critic say?

Invite opposition. Strong thinking survives pressure.

What would a different worldview see?

Break the tunnel. Let another frame challenge your own.

What is the human consequence?

Ask who may be affected if this output is wrong, shallow, biased, manipulative, or incomplete.

What must remain mine?

The final judgment. The moral responsibility. The taste. The meaning. The conviction.

This checklist is not bureaucracy.

It is cognitive self-defense.

FROM PEANUTS TO PARADIGMS

The full journey of this book can now be seen clearly.

The Peanut Contract begins with a small surrender:

"Let the machine do the thinking."

The flimsy mind accepts because thinking is uncomfortable.

The lethargic user accepts because convenience feels like productivity.

The hallucination fog hides the danger because the output looks intelligent.

Cognitive offloading deepens the dependency.
The status quo machine repeats the past and calls it wisdom.
The impossible lens breaks the frame.
The verification ritual restores truth.
Provocation replaces prompting.
The stewardship mind restores hierarchy.
The Paradigm Studio replaces the Peanut Factory.
The Peanut Economy is exposed as the wider system that cheapens thought.
The ethics of mental ownership restores responsibility.
And now the Human Upgrade turns the whole journey into action.
This is the movement:
From convenience to courage.
From output to insight.
From answers to questions.
From imitation to originality.
From speed to discernment.
From borrowed language to owned conviction.
From the Peanut Factory to the Paradigm Studio.
From cognitive surrender to sovereign mind.
That is the arc.
That is the upgrade.

THE CALL TO ACTION

So what must be done?
Not someday.
Now.
Students must stop using AI to escape learning and start using it to strengthen intellectual stamina.
Designers must stop asking for average solutions and start using AI to expose the assumptions that make industries boring.
Leaders must stop hiding behind dashboards and models and start standing visibly inside responsibility.
Teachers must stop grading polish alone and start cultivating friction, verification, and judgment.
Creators must stop generating endless variations and start choos-

ing what carries life.
Ordinary people must stop letting platforms rent their attention and start rebuilding the inner space where thought can breathe.
Organizations must stop measuring only output and start rewarding insight.
Schools must stop treating AI as only a cheating problem and start treating it as a judgment problem.
Societies must stop confusing automation with wisdom.
And each person must make a private decision:
Will I use AI to become sharper, deeper, freer, more responsible, more imaginative?
Or will I use it to become faster, shallower, more dependent, more fluent, and less awake?
That is the choice.
No software update will make it for you.

THE LAST REFUSAL

The Status Quo Machine will continue whispering.
It will say:
"Do not struggle."
"Do not read deeply."
"Do not verify."
"Do not sit in uncertainty."
"Do not risk originality."
"Do not take responsibility."
"Let me smooth the world for you."
"Let me make you comfortable."
"Let me finish your thought before you have begun it."
This is the voice of the Peanut Factory.
It does not sound evil.
It sounds helpful.
That is why it is dangerous.
The upgraded human must learn the sacred refusal:
No.
I will not sell my brain for peanuts.
I will not trade judgment for convenience.
I will not mistake generated fluency for earned insight.
I will not outsource my conscience.

I will not let the machine become the author of my convictions.
I will not become a clerk in the factory of average thought.
This refusal is not anti-technology.
It is pro-soul.
It is the refusal to let the most powerful tools in history shrink the human beings who use them.

THE FINAL HUMAN MANIFESTO

I will use AI, but I will not obey it.
I will learn from AI, but I will not kneel before it.
I will let it draft, summarize, compare, simulate, organize, and provoke.
But I will reserve for myself the sacred duties of judgment, conscience, meaning, taste, responsibility, and imagination.
I will not confuse speed with intelligence.
I will not confuse polish with truth.
I will not confuse output with insight.
I will not confuse convenience with freedom.
I will not confuse probability with possibility.
I will read what matters.
I will verify what matters.
I will question what matters.
I will create what matters.
I will stand behind what I release into the world.
I will use the machine as apprentice, mirror, critic, and catalyst.
But I will remain the steward.
I will not be merely an answer-seeker.

I will become an answer-questioner.
I will not be a tenant in my own mind.
I will become sovereign.
The future belongs not to those who get answers faster.
It belongs to those who question answers better.

CLOSING: TAKE FLIGHT

The factory doors are open.
Behind you, the conveyor belts continue to move. They will not stop. The Peanut Economy will continue producing its endless stream of summaries, slogans, dashboards, templates, automated essays, synthetic opinions, and beautifully packaged mediocrity.
Let it.

You do not have to live there.
You can walk out.
You can build the studio.
You can protect the silence.
You can ask the harder question.
You can read the source.
You can challenge the frame.
You can use the machine without surrendering to it.
You can bring back the whole human mind: analytical, imaginative, ethical, embodied, rebellious, compassionate, and alive.

The sky above the Peanut Factory is still open.
The machine may calculate the probable.
But the human can still choose the necessary.
The machine may generate the answer.
But the human can still ask the question that changes the world.
So take back the crown.
Sharpen the question.
Verify the truth.
Own the consequence.
Create the meaning.
And do not sell your brain for peanuts.
Take flight.

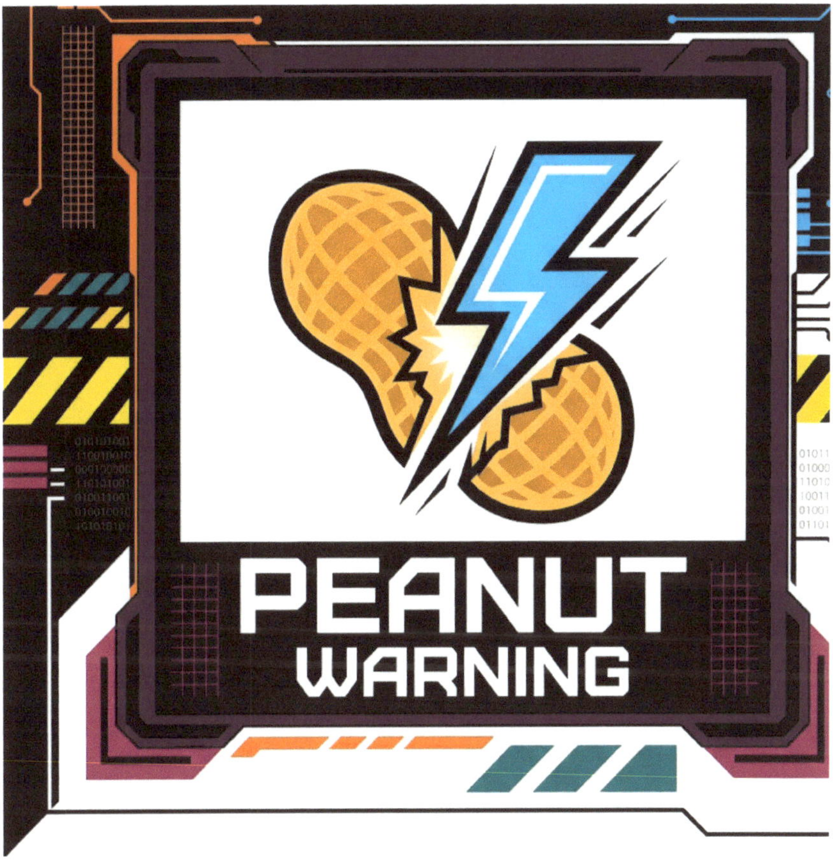
PEANUT
WARNING

Conclusion Don’t Sell Your Brain for Peanuts

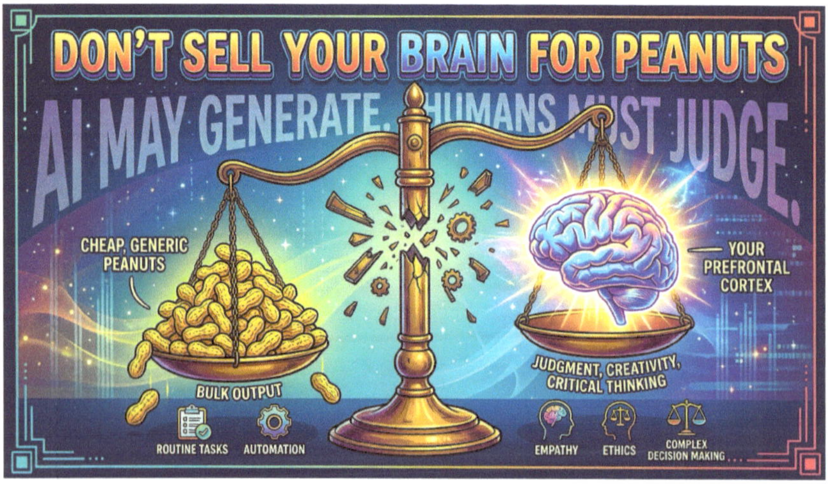

The Final Call to Think, Design, and Redesign Reality

We have arrived where this book first began: at the bargaining table of the modern mind.

On one side of the table sits convenience.

It is smooth, fast, polished, and persuasive. It promises to remove friction from your life. It says you no longer need to struggle with the blank page, wrestle with difficult books, sit with uncertainty, verify uncomfortable facts, or carry the burden of forming your own convictions.

It says:

"Let the machine think for you.
Let the platform decide what matters.
Let the summary replace the source.
Let the output replace the insight.
Let the answer replace the question."

And in exchange, it offers peanuts.

A little speed.
A little comfort.
A little productivity theater.
A little social approval.
A little polished mediocrity.
A little relief from the exhausting responsibility of being fully human.

On the other side of the table sits your brain.

NOT AS A LUMP OF BIOLOGICAL TISSUE.
NOT AS A STORAGE DEVICE.
NOT AS A CALCULATOR.
NOT AS A MACHINE WAITING TO BE MADE OBSOLETE.

Your brain is a living instrument of judgment, imagination, ethics, memory, intuition, design, responsibility, and transformation. It is the place where reality is not merely received, but questioned. It is the place where the obvious can be challenged, the inherited can be redesigned, and the impossible can become a new design brief.

Your brain is the organ of refusal.

It can look at a system and say, "This is not enough."
It can look at a polished answer and say, "This is not true yet."
It can look at a convenient shortcut and say, "This will make me weaker."
It can look at a machine-generated output and say, "Useful, but not final."
It can look at the status quo and say, "Who decided this was permanent?"
That power is priceless.
Do not sell it for peanuts.

THE JOURNEY WE HAVE TAKEN

This book began with a warning: human beings are in danger of trading cognitive sovereignty for cheap convenience.

We saw how the flimsy mind avoids critical thinking because truth is often uncomfortable. We saw how lethargy disguises itself as practicality, how "good enough" becomes a disease, and how people become spectators of reality instead of redesigners of it. We examined the hallucination fog, where AI can produce beautiful sentences without truth beneath them. We confronted the lethargic user who blames the machine while secretly enjoying the chance not to think. We looked at cognitive offloading, where useful tools become dangerous when they replace unused mental muscles.

We saw how AI can become a status quo machine, repeating the past with impressive fluency. We explored the designer's impossible lens, where "impossible" is often not a fact, but a badly framed question. We turned prompting into provocation, insisting that a prompt should not merely request an answer, but awaken a deeper line of inquiry.

Then we reclaimed judgment.

We named the Stewardship Mind: the human posture that treats AI as apprentice, mirror, critic, and catalyst, never as master. We asked whether AI will create a generation of deeper thinkers or a peanut factory of instant mediocrity. We exposed the Peanut Economy, where platforms, speed, viral summaries, shallow expertise, and productivity theater reward output instead of insight. Finally, we entered the ethical heart of the matter: mental ownership.

Who owns a thought generated by a machine?

Who is responsible when an AI-assisted decision harms someone?
What happens when people stop forming their own convictions?
The answer became unavoidable:
You can outsource a task,
but you cannot outsource responsibility.
And from there, the Human Upgrade emerged.
Not a technological upgrade.
A human one.
A call to become sharper, slower when necessary, braver in questioning, more disciplined in verification, more original in imagination, and more accountable in action.

AI WILL CONTINUE TO EVOLVE

Artificial Intelligence will not slow down for our comfort.
The models will become faster.
The outputs will become smoother.
The voices will become more natural.
The images will become more convincing.
The assistants will become more integrated into work, school, creativity, government, business, and daily life.
The machine will become easier to trust precisely because it will become easier to use.
That is why the danger will not always look dangerous.
The deepest threat may not arrive as a monster.
It may arrive as a helpful assistant.
A friendly interface.
A perfect summary.
A clean recommendation.
A confident answer.
A beautifully formatted hallucination.
A decision that feels objective because it came through a system.
This is why the human task remains unchanged.
No matter how advanced the machine becomes, the human must still think deeply, question bravely, verify carefully, imagine boldly, and redesign what appears fixed.
The tools will change.
The responsibility will not.

THE TWO FUTURES

There are two futures available to us.

The first is the Peanut Factory.

In this future, humans use AI to avoid effort. Students submit without learning. Designers generate without seeing. Leaders decide without owning consequence. Teachers grade polish instead of process. Creators drown in variations without vision. Platforms reward speed, outrage, and shallow certainty. Workplaces measure motion and mistake it for meaning.

Everything becomes faster.

Everything becomes easier.

Everything becomes thinner.

The world fills with content, but loses insight.

It fills with answers, but loses questions.

It fills with confidence, but loses truth.

That future is possible.

It is already being built wherever humans use powerful tools to deepen laziness instead of deepen thought.

The second future is the Paradigm Studio.

In this future, humans use AI to think harder. Students use it as a sparring partner. Designers use it to expose clichés and break frames. Leaders use it to test assumptions while remaining accountable. Teachers use it to train judgment. Creators use it to provoke originality. Ordinary people use it to verify, compare, question, and strengthen their own minds.

Here, AI does not replace human intelligence.

It pressures it.

It stretches it.

It irritates it into growth.

It becomes a lever, not a substitute.

A catalyst, not a conclusion.

A mirror, not a master.

That future is also possible.

But it will not happen automatically.

It must be chosen.

THE FINAL RULE

So let this be the permanent rule of the age:
Do not use AI to escape thinking.
Use AI to deepen thinking.
Use it to test hidden assumptions.
Use it to compare worldviews.
Use it to simulate futures.
Use it to prototype impossible ideas.
Use it to reveal contradictions.
Use it to accelerate research without killing understanding.
Use it to expand creative options without surrendering taste.
Use it to challenge your first answer.
Use it to make your thinking less lazy, not more dependent.
But do not use it as a hiding place.

DO NOT USE IT TO AVOID READING.
DO NOT USE IT TO AVOID CHECKING.
DO NOT USE IT TO AVOID MORAL RESPONSIBILITY.
DO NOT USE IT TO MANUFACTURE CONVICTIONS YOU HAVE NOT EARNED.
DO NOT USE IT TO COVER EMPTINESS WITH POLISH.
DO NOT USE IT TO RENT OUT YOUR MIND.

THE FINAL CALL

The future will not belong to those who merely get answers faster.
Speed will become common.
Output will become cheap.
Polish will become ordinary.
The future will belong to those who question answers better.
Those who can ask:
What is missing here?
What assumption is hiding underneath this?
What would disprove this?
Who benefits if I believe this?
Who is harmed if this is wrong?
What frame is this answer trapped inside?
What possibility is being excluded because it looks impossible?
What must remain human here?
These are not decorative questions.
They are survival questions.
They are design questions.
They are ethical questions.

They are the questions that keep the human mind from becoming a rented room in someone else's machine.

DO NOT SELL THE PRICELESS

Your mind is not a cheap input.
Your attention is not disposable.
Your judgment is not an outdated feature.
Your imagination is not a decorative luxury.
Your conscience is not a workflow inefficiency.
Your capacity to question, doubt, verify, imagine, and redesign is not a minor human inconvenience waiting to be automated away.
It is the very force that makes human beings capable of changing reality.
So do not accept peanuts for it.
Do not accept shallow productivity in exchange for deep perception.
Do not accept instant certainty in exchange for earned conviction.
Do not accept automated polish in exchange for lived meaning.
Do not accept platform visibility in exchange for inner sovereignty.
Do not accept the comfort of the factory when you were born with the power to build the studio.
The world does not need more fluent surrender.
It needs awake minds.
It needs designers of better questions.
It needs leaders who own consequence.
It needs teachers who protect friction.
It needs students who learn the grappling.
It needs creators who hunt anomalies.
It needs ordinary people brave enough to pause before believing, sharing, judging, and obeying.
It needs humans who refuse to become clerks of the probable.

FINAL WORDS

The machine will continue to generate.
Let it.
The platforms will continue to accelerate.
Let them.
The Peanut Economy will continue to reward noise.
Let it expose itself.

Your task is different.
Think deeply.
Question bravely.
Verify carefully.
Imagine boldly.
Design courageously.
Redesign reality where it has become lazy, unjust, shallow, or false.
Use AI, but do not kneel before it.
Learn from it, but do not disappear into it.
Let it assist your mind, but never let it own your judgment.
The final command is simple:
Do not use AI to escape thinking.
Use AI to deepen thinking.
Do not accept peanuts for the priceless force of your mind.
Tear up the Peanut Contract.
Walk out of the factory.
Build the Paradigm Studio.
Take back the crown.
And never sell your brain for peanuts.

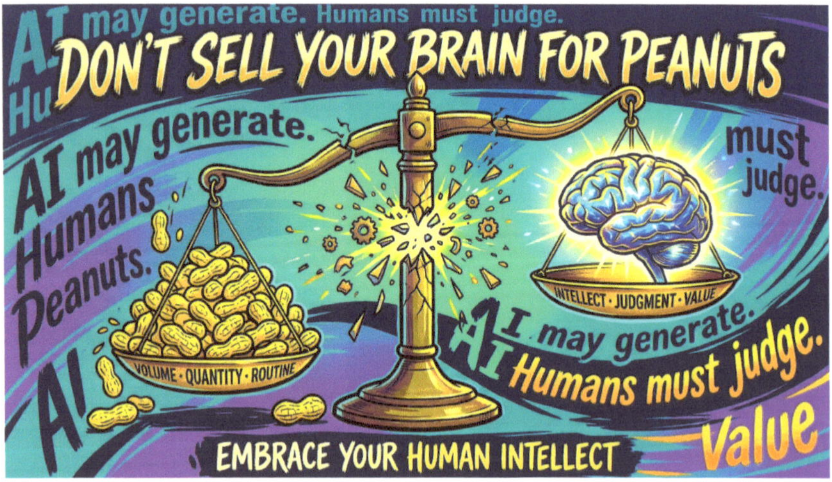
AI may generate. Humans must judge.
DON'T SELL YOUR BRAIN FOR PEANUTS
AI may generate.
Humans
Peanuts.
AI
must
judge.
VOLUME · QUANTITY · ROUTINE
INTELLECT · JUDGMENT · VALUE
AI may generate.
AI Humans must judge.
Value
EMBRACE YOUR HUMAN INTELLECT

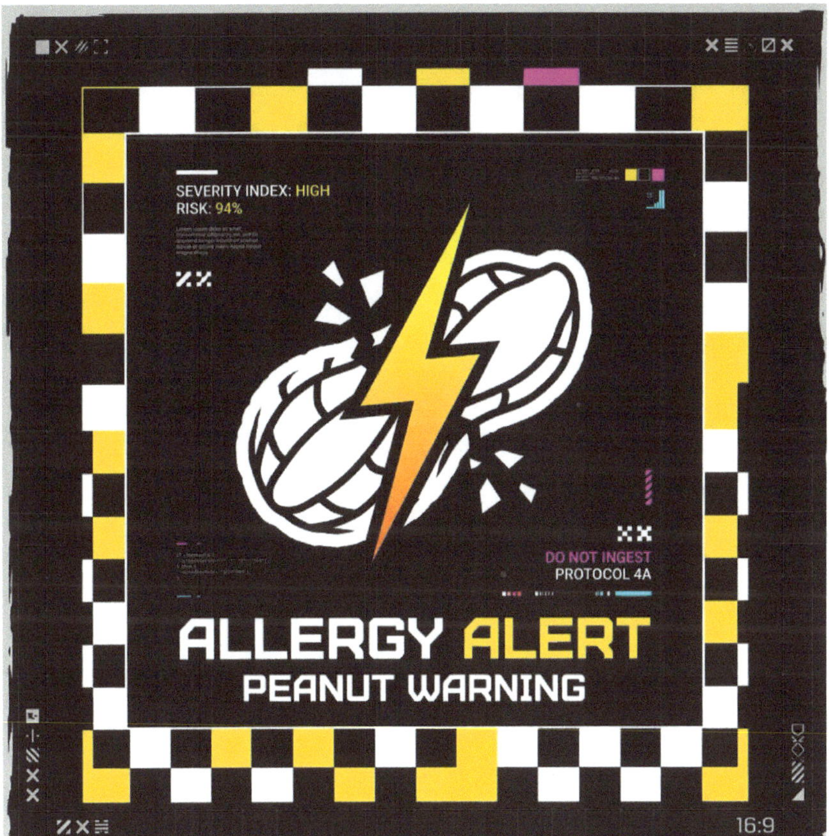
SEVERITY INDEX: HIGH
RISK: 94%
DO NOT INGEST
PROTOCOL 4A
ALLERGY ALERT
PEANUT WARNING
16:9

Don't Sell Your Brain for Peanuts is a sharp, poetic, and practical critique of mental laziness in the age of artificial intelligence. It asks a provocative question: are we using AI to expand human intelligence, or are we quietly outsourcing our ability to think?

Through the metaphor of "peanuts," the book exposes how people trade deep thinking for instant answers, confuse fluency with truth, and allow hallucinated confidence to replace careful reasoning. But this is not an anti-AI book. It is a book about human stewardship.

For designers, educators, creators, students, and leaders, this book offers a new discipline: use AI not as an oracle, but as a mirror, critic, apprentice, and catalyst. The goal is not faster mediocrity. The goal is deeper originality.

Because the future does not need more copied answers.

It needs people brave enough to think.

Thankyou for reading this far!

www.ingramcontent.com/pod-product-compliance
Lightning Source LLC
LaVergne TN
LVHW052149210926
848482LV00019B/299
* 9 7 9 8 9 0 5 1 0 2 2 0 2 *